Renewable Energy: Today's Contribution, Tomorrow's Promise

Cynthia Pollock Shea

Worldwatch Paper 81
January 1988

Library of Congress Catalog Card Number 87-51638
ISBN 0-916468-82-8

Printed on recycled paper

Table of Contents

When faced with a severe wood shortage due to overcutting, the ancient Greeks began to design their homes and cities so as to take advantage of the sun's ability to warm buildings in winter and cool them in summer. Through the ages many civilizations have learned to tap the multiple processes set in motion by the sun to garner useful energy. Uneven heating of the earth's surface produces wind, yesterday's winds are today's waves, and a season's worth of solar energy is trapped in plants whose residues can be burned as fuel.

Almost 15 years after the first major oil price rise wreaked havoc on the world's economies, efforts to tap the planet's myriad sources of renewable energy have met with mixed success. The 1986 oil price collapse further set back many renewables programs, particularly in the United States. Renewables offer a timely alternative to dwindling oil supplies and to environmentally damaging coal combustion, but policy support and financial backing need to be strengthened if their contribution is to increase significantly in the years ahead.

Expanded use of renewables and a greater commitment to energy efficiency are the most cost-effective and environmentally sound approaches to mitigating many seemingly intractable problems. In the United States, fossil fuel pollutants may cause as many as 50,000 premature deaths annually. Across the Atlantic, the air pollution caused by burning these fuels is implicated in damaging 31 million hectares of trees in central and northern Europe. Each year fossil fuel combustion emits some 5.4 billion tons of carbon. Forest clearing releases another 1.0–2.6 billion tons per year. Atmospheric carbon dioxide concentrations have increased 9 percent since 1960 and 30 percent since 1860, contributing to the greenhouse effect that is predicted to change the earth's climate.[1]

I sincerely thank Hilary French for invaluable research assistance and Susan Norris for production assistance. I am also grateful to Terry Clausen, Willem Floor, Paul Gipe, George Hagerman, Tom Jaras, Mike Jones, John Kadysewski, Eric Larson, Don Marier, Paul Maycock, Ross Pumfrey, and especially Kevin Porter for reviewing early drafts of this manuscript.

Completely averting global warming no longer appears possible. But strengthening the commitment to renewable energy sources can help delay the warming, as well as the rise in sea level and changed precipitation patterns that would accompany it. Many renewables technologies do not require combustion, while those that do will not be net carbon contributors if enough planting occurs to offset the amount of biomass burned. Hydropower alone effectively displaces 578 million tons of carbon emissions from coal-fired power plants each year, equivalent to over 10 percent of total 1987 emissions from fossil fuels.[2]

Biomass and hydropower, two long-used renewable energy sources, already provide about 21 percent of the total energy consumed worldwide, 15 percent and 6 percent respectively. The distribution of this use is uneven: Some of the poorest developing countries derive more than 75 percent of their energy from the direct combustion of biomass; others, well-endowed with water resources, obtain most of their electricity from hydro projects.[3]

The literature is replete, unfortunately, with projects that failed when they exceeded the renewable capacity of natural systems. An energy source is only renewable if, with proper management, its sustained use will not deplete supplies. Wind, sunlight, flowing water, plants, and forests are examples of seemingly perpetual energy sources. In many areas where fuelwood is the major contributor to energy supplies, however, it can no longer be thought of as renewable. Consumption rates have overtaken sustainable yields. An early calculation of environmental limits is thus an essential component of any renewable energy project.

Despite erratic political and financial support, some of the new renewable energy technologies have made remarkable strides during the past decade. Investments in their development and use now total some $30 billion annually. Two-thirds is spent on hydroelectric projects financed by leading banks. Some sectors—small hydropower, geothermal energy, and biomass—are growing rapidly. Others, such as the wind turbine and solar thermal equipment industries, are

"Biomass and hydropower already provide about 21 percent of the total energy consumed worldwide."

going through a difficult shakeout period during which marginal companies will fold.[4]

A few countries also appear to be moving in the right direction. Brazil, Israel, Japan, the Philippines, and Sweden are well on their way toward major reliance on renewable energy sources. Brazil, for example, obtains almost 60 percent of its energy from renewables. In 1986, Brazil was the world's largest producer of alcohol fuels, the second largest producer of fuelwood, and the fourth largest producer of hydroelectric power. Even the most modern sectors of the economy, such as automobiles and steel production, depend on these energy sources.[5]

Brazil illustrates well both the pitfalls and the promise of reliance on renewables. Many of its large hydro projects were approved without rigorous environmental assessments and without first introducing energy efficiency measures. In some cases people were uprooted and unique plant and animal species eradicated only to build generating capacity that went unused. Other programs have resulted in real progress. One-third of the steel manufactured in the country relies on charcoal for the smelting process. Fast-growing eucalyptus plantations provide most of the wood. Higher biomass yields and more efficient charcoal making and smelting processes will soon mean that just one-fifth as much land is required to support a given level of steel production as was required in the seventies.[6]

In most other countries, efforts on renewables have ebbed and flowed with the political tide. A concerted effort to increase their use will help head off impending environmental crises and the threat posed to national security by unreliable fuel imports. To take advantage of the growth spurt in renewables that is sure to accompany the next upward climb on the oil-price roller coaster, countries need to lay the institutional groundwork today.

Removing price subsidies for fossil fuels, for example, would enable renewables to compete fairly in the marketplace. Increasing the funds available for research and development (R&D), or diverting them from other areas, would spur the advance of promising new technologies. And enhanced international cooperation in R&D would cut costs and facilitate the spread of improved renewable energy technologies. If expanded to include developing countries, such cooperation would aid those economies most in need of restructuring their energy sectors and wean them from imported petroleum. Developing countries typically do not have the money, or the expertise, to independently launch major renewable energy programs.

Much of the Third World can take special advantage of new renewables technologies, however, because most are small in scale, have zero or minimal fuel costs, and can often be assembled with local labor. Small-scale technologies with short construction times provide greater adaptability in responding to unpredictable growth in power demand. Economies that depend on renewable fuels are not as vulnerable to supply disruptions or price volatility, nor are they forced to spend their foreign exchange on fuel imports. Half of all developing countries rely on imported oil for over 75 percent of their commercial energy needs. Yet sunshine, wind, water, and biomass are all available locally.[7]

An intensified global commitment to renewable energy sources will put the world economy on more stable footing. Building resilience into energy policies via efficiency measures and diversified, smaller scale supply options will help provide the flexibility needed to adapt to an unpredictable future. Unfortunately, many energy policymakers, complacent after the oil price plunge of the mid-eighties, are not looking ahead.

Formulating an energy strategy that can sustainably meet future needs will require better analyses of the trade-offs among various energy sources. Energy planners and policymakers need to look beyond construction, fuel, and operating costs and assess variables such as the relative amounts of land and water required, the amount of carbon dioxide and other gases emitted, and the toxicity and volume of the waste products. They might ask, for example, which would inundate more land: large, new hydroelectric projects or expanded use of carbon-emitting coal combustion. New supplies must

"Small-scale technologies with short construction times provide greater adaptability in responding to unpredictable growth in power demand."

not only be appropriate to local circumstances, they must also help ward off threats to the global environment.

Harnessing Water's Power

Hydroelectric power supplied 21 percent of the world's electricity in 1986—less than coal and oil, but more than nuclear power. Growing interest in developing hydroelectric power is largely an outgrowth of governments' desire to be more self-sufficient in energy and to provide low-cost electricity, often to areas or industries not previously served by electric utilities. Initially, planners conducted surveys of potential sites for large-scale hydropower projects; more recently, attention has focused on the availability of sites for smaller facilities.[8]

In 1986, Venezuela completed the Guri dam, the largest in the world. With a 10,000-megawatt generating capacity it can produce as much electricity as 10 large nuclear power plants. Brazil is in the process of building a hydroelectric plant with 20 percent more capacity than the Guri, and China is contemplating an even larger one. At the same time, many countries, particularly in the Third World, are installing generators thousands of times smaller on remote rivers and streams. (A generating plant is usually classified as "small" if it has a capacity of 15 megawatts or less.) The electricity is used to provide power to isolated, sparsely populated communities and agricultural processing plants far from electric utility power lines.[9]

Most hydro plants larger than 1,000 megawatts that are under construction or planned are located in developing countries or in remote areas of industrial nations. Industrial countries have already tapped their most promising sites—areas with a steep, narrow gorge through which water can fall—and those locations that remain have generally been set aside as parkland or effectively excluded from consideration because of their natural beauty. Whereas North America and Europe had developed 59 and 36 percent of their large-scale hydropower potential, respectively, by 1980, Asia had harnessed just 9 percent, Latin America 8 percent, and Africa 5 percent.[10]

Brazil and China presently have the largest and most ambitious programs. China alone has 15,000 megawatts of large hydro projects under construction and plans to complete twice that much by the turn of the century. Brazil nearly tripled its hydroelectric generating capacity by adding 21,535 megawatts between 1973 and 1983; projects to tap the vast potential that remains are moving forward. New sites are further from population centers and would thus incur higher transmission costs, but enthusiasm still runs high in some quarters.[11]

Many who have studied hydropower developments in Brazil believe that the construction program has developed a life of its own and that more rivers are being dammed than necessary. Indeed, many of the dams now under construction are legacies of the military government, and would not be approved under today's stricter environmental regulations. The Balbina, for example, was ordered more than a decade ago, is not yet operational, and will flood an area of 1,554 square kilometers—half the size of Long Island and as much as required at another Brazilian dam (the Tucurui) that produces 15 times as much power. Since Brazil is fairly flat, its hydroelectric reservoirs are disproportionately large, and flooding them threatens many tropical species with extinction. And wildlife rescue efforts, poor as they may be, are better organized and funded than initiatives to resettle the human inhabitants.[12]

These monster dams constitute some of the largest engineering projects in the world. The 12,600-megawatt Itaipu dam in Brazil is five miles long, and half as high as the Empire State Building. World Bank projections indicate that 223,560 megawatts of large hydro capacity will be added in developing countries between 1981 and 1995, more than half of it in Brazil, China, and India. This is equivalent to 225 large nuclear plants, or 82 percent of the world's nuclear capacity in 1986. Thirteen developing countries collectively installed more than 40,000 megawatts of hydroelectric power between 1980 and 1985. (See Table 1.) But future growth is not likely to be as robust as anticipated because of the Third World debt crisis and the consequent shortage of investment capital.[13]

Table 1: Thirteen Largest Additions to Hydroelectric Capacity in Developing Countries, 1980–85

Country	Operating Capacity 1980	Operating Capacity 1985	Increase
		(megawatts)	
Brazil	27,267	42,762	15,495
China	20,318	25,788	5,470
Colombia	2,908	5,939	3,031
Romania	3,414	5,914	2,500
India	11,794	14,211	2,417
Mexico	6,491	8,626	2,135
Yugoslavia	6,115	7,841	1,726
Vietnam	330	1,800	1,470
Turkey	2,131	3,575	1,444
Pakistan	1,800	3,200	1,400
Zaire	1,077	2,477	1,400
Philippines	940	2,195	1,255
Nigeria	760	1,900	1,140
Total	85,345	126,228	40,883

Source: World Bank, *A Survey of the Future Role of Hydroelectric Power in 100 Developing Countries* (Washington, D.C.: 1984).

In the United States, which has the largest installed hydroelectric capacity, the dam-building era is winding to a close. Not a single new large dam was approved to receive federal funding between 1976 and 1986. For dams that receive federal dollars after 1986, local governments have to put up half the money, a requirement that likely will cause many projects to be cancelled and others to be reduced in size. After 85 years of building massive water projects in the western United States, the U.S. Bureau of Reclamation announced in 1987 that its mandate to tap new water supplies had virtually expired and that the agency would halve its work force over the next decade.[14]

Any large new supplies of hydroelectric power for the United States are likely to be imported from Canada. In 1986, transborder sales totaled 12.7 billion kilowatt-hours, about 0.5 percent of U.S. electricity use. During the nineties, New Englanders may receive 7 percent of their electricity from Quebec, for a price of about $3 billion. Offers to sell additional hydroelectric power and to build new dams dedicated in part to exporting power have been made by British Columbia, Quebec, and Manitoba. The transactions hold special appeal to northeastern states with high electricity prices and expensive nuclear plants.[15]

In regions of the world where water is scarcer, plans to build dams and reservoirs can lead to international battles. Turkey, for example, intends to vastly expand its hydroelectric capacity, but construction of its Ataturk dam on the Euphrates has already lowered water levels downstream. Five of the eight 100-megawatt turbines at Syria's Tabaq dam, completed in 1978, are reportedly out of the water as a result. Iraq claims the same injustice has occurred to it from the Syrian dams. Yet hydropower is a major component in the electricity supply systems of Egypt, Lebanon, and Syria, providing over 40 percent of total capacity, and expansion plans are still being drawn up.[16]

Huge dams can make substantial contributions to economic development in electricity-short countries, but like any large electricity generating option, there are trade-offs. Reservoirs inundate forests, farmland, and wildlife habitat, and uproot entire communities of indigenous peoples. If China proceeds with its Three Gorges project—the world's largest at 13,000 megawatts—several million people will be displaced. An additional million people will be forced from their homes in central India if a project to build 3,000 dams in the Narmada Valley is pursued.[17]

Impounding a river radically changes the surrounding ecosystem. Nutrient-bearing sediments, instead of being deposited on agricultural floodplains and providing food for downstream fish, accumulate behind turbines and dams. Hydroelectric dams may also change the temperature and oxygen content of downstream waters, altering the mix of aquatic and riparian species. The construction of ever higher dams—113 will exceed 150 meters by 1990—and the increased prevalence of seismic activity near reservoirs are leading many to speculate that the combination of increased water pressure and unstable geological formations will result in more frequent and severe earthquakes. In tropical environments, reservoirs also expand the breeding grounds for the carriers of malaria, schistosomiasis, and river blindness.[18]

The reservoirs behind many large dams, especially those downstream from deforested and poorly managed watersheds, have silted up faster than anticipated, substantially shortening the working life of projects and jeopardizing their economic viability. Siltation behind the Ambuklao dam in the Philippines is estimated to have cut the project's useful life in half. Revenue losses caused by sedimentation behind the Cachi dam in Costa Rica have been estimated to reach between $133 million and $274 million over a 50-year period. In Colombia, an innovative program to transfer financial resources from the lowland beneficiaries of hydropower development to upland farmers is under way. A sales tax on electric power from major hydroelectric plants has been authorized to help stabilize upland watersheds through soil conservation and reforestation.[19]

Programs to alleviate the negative effects of dams on fish populations include the installation of screens to divert fish away from the turbines' dangerous blades and ladders to assist fish swimming upstream. More careful placement of intake structures can reduce downstream temperature changes, and increased turbulence raises the water's oxygen level. These mitigation strategies have met with mixed success and their effectiveness or lack thereof is the primary factor blocking continued expansion of hydropower in the northwestern and northeastern United States.[20]

On a smaller scale, hydropower generation not connected to a central grid totaled nearly 10,000 megawatts worldwide in 1983. Most of these projects were in developing countries, but West Germany alone had 3,000 small units in operation. Frost & Sullivan, a market research firm, predicts that decentralized hydropower in stand-alone systems will climb to 36,000 megawatts by 1991.[21]

A World Bank survey of 100 developing countries found that 31 of them had more than doubled their hydroelectric capacity between 1980 and 1985. At least 28 of these have small-hydropower programs. In Burundi, Costa Rica, Guatemala, Guinea, Madagascar, Nepal, Papua New Guinea, and Peru, small-hydropower potential exceeds total installed generating capacity from all energy sources. China leads the world in this field, with some 90,000 turbines supplying electricity to rural areas.[22]

Industrial countries are also realizing the contribution that small hydro projects can make. By 1985, private entrepreneurs in the United States had brought almost 1,000 megawatts of small hydropower on-line and electric utilities had installed more than twice that amount. Almost 60 percent of the total 3,200 megawatts came on-line during the eighties. Elsewhere, small dams that had fallen into disrepair are also being refurbished. Poland has begun to rehabilitate 640 small plants and in the Canadian province of Ontario, 570 formerly developed sites have been identified.[23]

The technology for generating electricity from falling water is well developed. Commercial systems have been available since the turn of the century, when the first hydroelectric plant started producing power at Niagara Falls on the U.S.-Canadian border. With most manufacturers of turbine technology located in Europe and North America, however, hydro development in the Third World is often dependent on foreign equipment suppliers.

Hydro projects are capital-intensive, and foreign currency requirements can mount rapidly. The capital costs of large projects are generally in the $500–2,500 per kilowatt range, while those for small projects vary from less than $1,000 to over $6,000 per kilowatt. Average costs for small facilities hover around $2,000. The primary factors that determine project cost are the design, the equipment used, and the frequent need for new civil works. Limiting the tendency to overengineer projects can save several thousand dollars per kilowatt. To this end, China, Colombia, India, Indonesia, Nepal, Pakistan, and Thailand have all developed the capability to manufacture small tur-

"To operate well for many decades, hydro projects require sound management of entire watersheds."

bines domestically. China also exports these turbogenerators to both industrial and developing countries.[24]

To operate well for many decades, hydro projects require sound management, not just of equipment, but of entire watersheds. Fragmented institutional structures impede enlightened management because each function of a watershed belongs to a different agency. As Brandeis University Professor Donald Worster writes in *Wilderness* magazine, "Everybody wants a piece of [rivers], wants to siphon them off, dump wastes in them, drink from them, or move barges along them, but no one has ever been given overall charge of protecting their renewability."[25] Hydroelectric power will not be truly renewable until the functions of flood control, irrigation, transportation, power production, tree planting, fisheries management, and sanitation are coordinated within the overall goal of maintaining healthy and productive rivers.

The power of the oceans is also being investigated as a potential source of electricity. Many experimental and a few commercial facilities have been developed to capture the energy of tides, waves, and the temperature difference between surface and deep waters.

Although small tidal mills to provide mechanical power were common along the north Atlantic coasts of Europe and North America hundreds of years ago, only in recent decades has interest in using tides to generate electricity been rekindled. Numerous small and several fairly large plants are now operating. Each relies on a dam built across the mouth of a cove to form a pond. As the rising tide comes in, sluice gates open to allow water to enter. Soon after high tide the gates are closed and water returning to the sea is directed through an electricity generating turbine.

The power available from a tidal pond is proportional to the square of the tide's range: A three-meter tide generates nine times as much power as a one-meter tide. A survey of proposed projects worldwide indicates that a minimum tidal range of three to five meters is required for a tidal power plant to be considered economically feasible.

The world's greatest resource exists at the upper end of the Bay of Fundy in Canada, where the tidal range reaches 10–12 meters.[26]

For two decades the world's largest tidal plant, a 240-megawatt unit, has performed well at the La Rance estuary in northern France. It was built as a prototype, but is now operated commercially. A small 0.4-megawatt plant started operating near Murmansk in the Soviet Union at the same time. The Soviets plan to build a 40-megawatt plant at Mezen Bay on the Kola Peninsula and a 300-megawatt unit at Lumbovka on the White Sea. China brought a 10-megawatt plant on-line in 1986 after gaining operating experience at several smaller facilities. An 18-megawatt tidal plant that was installed in an existing flood control dam started operating in 1984 at Annapolis Royal, Nova Scotia, on the Bay of Fundy.[27]

Canada is assessing the merits, and perhaps more importantly the drawbacks, of building 1,000-megawatt and larger facilities in the Bay of Fundy. Scientists are studying the environmental impacts that such large plants might have, both inside and outside the proposed tidal reservoirs. Potential impacts inside include reduced tidal range, which would dry out the perimeter, and reduced tidal current flows. Potential outside-basin effects include altered sea levels, perhaps as far south as Cape Cod, Massachusetts, and the death of migratory fish species.[28]

The United Kingdom is considering a proposal to dam the Severn Estuary on the west coast of England. The 16-kilometer barrage would cost 7 billion pounds ($4 billion), and 192 turbogenerators would be capable of producing 7,000 megawatts of electricity, enough to meet 5 percent of the country's current electricity demand early next century.[29]

A problem common to all such facilities is the cyclical nature of the tides coupled with the fact that the most economical tidal plants only produce electricity during ebb tide. Virtually all the requisite technology has been developed, for conventional hydroelectric projects and marine construction firms, but the large economies of scale asso-

"After years of experimentation, Norway brought the world's first wave power plants on-line in early 1986.

ciated with tidal projects require multibillion-dollar investments. Governments are looking to the private sector for financing, but such costly high-risk ventures for projects that would only produce power part of the time hold little appeal in today's energy markets. If backers could be found, they would surely demand government assistance or participation to mitigate the risks.[30]

After years of experimentation, Norway—a nation rich in hydropower, oil, and natural gas—brought the world's first wave power plants on-line in early 1986. The two prototype plants have a combined capacity of 0.85 megawatts. One plant consists of a 5,500-square-meter reservoir that is open to the sea via a 90-meter tapered channel that narrows as it moves into the reservoir. As waves approach the apex, they spill water over the channel sides, keeping the reservoir three meters above sea level. As it returns to the ocean, the water is directed through a turbine, generating electricity.

The second plant, called an oscillating water column, produces electricity from both the rise and fall of individual waves. When the water level rises in the column, it pushes air up through a turbine; as the water level falls, air is pulled in from the atmosphere. Once again, the air passes through a turbine, which spins in the same direction, regardless of which way the air is flowing. Built for less than $1,900 per kilowatt, the column technology generates power for 4–6¢ per kilowatt-hour—a competitive rate.[31]

The success of these prototypes and the ease with which prospective purchasers can survey the site, only 50 kilometers from the Bergen airport, have prompted Indonesia, Portugal, and Puerto Rico to place orders. Norway is also pleased with the results and plans to build larger plants by 1990. Ironically, the United Kingdom, which did much of the R&D work and then abandoned wave power as uneconomical, has now decided to build its own 200-kilowatt pilot plant on the island of Islay off Scotland's west coast, based on Norway's experience. The State Energy Commission of Western Australia has also decided to support the construction of a 1-megawatt plant at Esperance.[32]

Japan operates several hundred small wave generators (0.5-3.0 kilowatts) to power its navigational buoys and has tested larger turbines (125 kilowatts) aboard the experimental ship Kaimei in the Sea of Japan. Other wave generating technologies in the range of 1 to 20 kilowatts have been developed and tested in the Soviet Union, Sweden, and the United States.[33]

Temperature differences between the warm ocean surface and colder, deeper layers may someday be used to generate electricity. Covering 70 percent of the earth's surface, the ocean is by far the largest natural solar collector in the world. Surface temperatures in the tropics are typically 27 degrees Celsius, and the water at 1,000 meters or below is about 4 degrees Celsius. In a closed-cycle system, the warm water gives up its heat to a fluid with a low boiling point, most commonly ammonia or freon. After powering a turbogenerator, the steam is condensed by cold water from the depths.[34]

In an open-cycle system, seawater itself is boiled in a vacuum chamber, producing both electricity and desalinated water. Within the next five years, the U.S. Department of Energy plans to construct a 165-kilowatt prototype of an open-cycle facility off the coast of Hawaii. Accounting for seawater and vacuum pump requirements, the plant should have a net output of 40–50 kilowatts. In the future, small ocean thermal energy conversion facilities may be used by energy- and water-short islands.[35]

Of potential use in both oceans and rivers are turbines that harness the energy of water currents. Prototypes ranging in size from 20 to 100 kilowatts have been tested in Canada, Japan, and the United States. Like wind turbines, these come in both horizontal-axis and vertical-axis varieties. Commercial units based on these designs are considered economically feasible at flow speeds of one to two meters per second.[36]

New Uses for Biomass

Biomass, derived directly or indirectly from plant photosynthesis, is a versatile fuel source capable of providing high-quality gaseous, liquid, and solid fuels as well as electricity. Primary sources include forestry and wood processing residues, crop residues and animal wastes, and energy crops. Though less than 1 percent of annual biomass growth is tapped for energy, it provides 15 percent of the energy used worldwide.[37]

An important distinction to make when assessing biomass-based systems is whether the process converts wastes and residues into fuel, which increases the efficiency of existing economic activity, or whether it instead requires the cultivation of biomass specifically for energy, which requires all the inputs of an agricultural or silvicultural production system.

Wood is the most widely used biomass energy source. Over half the wood cut each year is burned to produce energy. According to U.N. statistics, the world's largest fuelwood producers, in order, are India, Brazil, China, Indonesia, the United States, and Nigeria.[38]

Available energy potential from forest and timber industry residues was found by the International Energy Agency (IEA) to be highest in the United States. (See Table 2.) With proper management, many countries can expand their fuelwood use without depleting natural forests. On average, about 25 percent of the wood entering the timber industry is available for conversion to energy.[39]

Many developing countries, on the other hand, are experiencing a severe wood shortage. Unprecedented and unsustainable rates of forest clearing to provide agricultural and grazing land, timber supplies, and fuel have left nearly 100 million people suffering from an acute scarcity of fuelwood, and 1.2 billion more with unsustainable supplies. Wherever cutting exceeds regrowth, wood is no longer a renewable fuel.[40]

In the United States, industrial, commercial, and utility applications account for some two-thirds of the wood used for energy. The remainder provides 10 percent of the nation's residential heating, warming 5.6 million homes exclusively and 21 million homes partially with wood. Almost half of all wood fuel is consumed by the pulp and

Table 2: Available and Recoverable Energy Potential from Agricultural, Forestry, and Timber Industry Residues, Selected Countries, 1979

Country	Available Potential from Forestry and Timber Industry	Recoverable Potential: Crop Residues	Recoverable Potential: Animal Wastes	Total	Share of Total Energy Requirements
	(million tons of oil equivalent)				(percent)
Turkey	5.9	5.4	1.5	12.8	41.7
Finland	8.6	0.4	0.2	9.2	36.8
Sweden	10.4	0.8	0.2	11.4	22.8
Canada	32.1	2.3	1.8	36.2	16.8
Austria	2.8	0.4	0.3	3.5	14.0
Spain	2.4	4.1	0.6	7.1	10.0
France	6.1	5.2	2.4	13.7	7.2
United States	68.5	20.2	5.0	93.7	5.2
West Germany	6.2	2.2	1.6	10.0	3.7
Japan	6.7	0.4	1.2	8.3	2.1

Source: International Energy Agency, *Renewable Sources of Energy* (Paris: Organisation for Economic Co-operation and Development, 1987).

paper industry, which meets more than 55 percent of its own energy needs. The second largest market is the lumber industry, and the fastest-growing sector is the non-forest-products industries. Users include electric utilities, industry giants such as Dow Corning and Proctor & Gamble, and companies formed specifically to take advantage of regional wood supplies by generating electricity and selling it to local utilities. (See Table 3.)[41]

Since 1983, four U.S. utilities have built wood-burning power plants, each able to generate more than 45 megawatts of electricity; their combined output is enough to supply some 175,000 homes. Accord-

Table 3: United States: Selected Biomass-Fueled Electricity Generating Facilities

Project	Capacity (megawatts)	Fuel Source	Start-Up
Forest Products Industry[1]			
Union Camp Corp. (VA)	96	pulping waste, peanut shells	1937
Champion Intl. Corp. (FL)	78	pulping waste, bark	1961
Manville Forest Prod. Co. (LA)	72	wood and pulping wastes	1961
Louisiana Pacific Corp. (CA)	26	wood waste	1983
Electric Utilities			
Northern States Power (WI)	72	forest residues[2]	1983
Burlington Electric Dept. (VT)	50	forest residues	1984
Eugene Water & Electric Board (OR)	46	mill residue	1983
Washington Water & Power (WA)	46	mill residues	1983
Independent Power Producers			
Ultrasystems (CA)	27	forest, industry and ag. residues	1988
Ultrasystems (ME)	27	forest residues	1986
Wheelabrator Energy (CA)	25	orchard prunings	1989
Alternative Energy Decisions (ME)	17	forest and industrial residues	1986
Nontraditional Producers			
The Lihue Plantation (HI)	26	bagasse	1980
Dow Corning (MI)	22	wood chips	1982
Farmers Rice Milling Co. (LA)	11	rice husks	1984
Proctor & Gamble (NY)	10	industrial waste wood, wood chips	1983

[1]Predominantly biomass-fueled. Sometimes supplemented with coal or natural gas.
[2]Multifuel capability, but has been burning mostly wood since 1983.

Source: Worldwatch Institute, based upon Meridian Corporation, *Electric Power From Biofuels: Planned and Existing Projects in the United States* (Washington, D.C.: U.S. Department of Energy, 1985), news reports, and private communications.

ing to a study conducted by the California Energy Commission, wood-fired boilers can be installed for about $1,340 per kilowatt, 20 percent less than a coal plant costs. In Oregon, Vermont, Washington, and Wisconsin, these utility customers are receiving the benefits of having their money stay in state, providing jobs and tax revenues instead of leaving the region or nation to pay foreign fuel suppliers.[42]

Industrial-sector fuelwood users are greater in number and spread more widely throughout the country. The largest market is in California, where almost two dozen 10–50 megawatt wood energy projects are on-line now and dozens more will be shortly. Over 500 megawatts of power are already on-line or under construction throughout the state. In Maine, a half-dozen such projects were scheduled for completion in 1987, with another half-dozen under construction. The more than 250 megawatts of wood energy projects in Maine, worth over $500 million, are equivalent to the rest of the economic development in the state combined.[43]

Leading U.S. forest products regions such as the Northwest and Southeast are ripe for expanded wood energy use, but they are not the only potential growth areas. States such as New York and Michigan also have vast forest resources, but regulatory hurdles have blocked their use.[44]

One assessment found that each year the state of Virginia produces enough sawdust, logging residue, and unsalable low-quality trees to replace 42 percent of the oil and gas consumed in its industrial and commercial sectors. Thus far, little of this potential has been harnessed, but industrial parks, colleges, hospitals, and a variety of other enterprises are likely to take the plunge once oil prices start creeping up again. Fuelwood costs are likely to remain low, especially if the supplier would otherwise be faced with mounting waste disposal charges.[45]

Wood-fired electricity is also being tried in smaller generating plants in the Third World. In the late seventies, 17 dendrothermal projects of 3 megawatts each were planned throughout the Philippines. Local farmers were to supply the fuel from plantations of leucaena, a fast-growing hardwood. Although British and French equipment for all 17 was delivered, only 4 plants were completed on schedule and just 2 are said to be operating reliably. The overly ambitious project was unable to garner sufficient financing, was blocked by bureaucratic infighting, and was based on excessively optimistic predictions of wood yields on marginal soils. A more modest effort might have succeeded.[46]

Crop residues and animal wastes are a significant by-product of the food-producing agricultural sector, but diverse agricultural policies, marginal energy conversion economics, and alternative, nonenergy uses make assessing their potential difficult. An IEA study found that using the most energy-efficient processes available in 1979, agricultural wastes could contribute between 0.4 and 8.2 percent of total energy requirements in IEA member countries, with the exception of Turkey, where the contribution could equal 22.5 percent. Denmark, Greece, Ireland, Portugal, and Spain were considered able to provide at least 5 percent of their total energy needs from crop residues and animal wastes.[47] In light of recent advances in biomass conversion technologies, a similar assessment conducted today would likely reveal even higher potential contributions.

In the tropical Caribbean, where sugar is the major export crop but producers face depressed world markets, more-efficient use of plant residues coupled with a switch to already tested fast-growing cane varieties that produce more biomass could significantly augment energy supplies. The amount of electricity available in Barbados, Cuba, the Dominican Republic, Guatemala, Guyana, and Honduras would increase severalfold. In Thailand, another sugar producer, cane residues could fuel 300 megawatts of electricity generating capacity and thereby add 25 percent to the value of the industry.[48]

The Hawaiian sugar industry started selling electricity in the late seventies and in 1985 supplied 58 percent of the power on the island of Kauai and 33 percent on the island of Hawaii. Sugar companies have installed at least 150 megawatts of capacity to burn bagasse, the residue after juice is extracted, and sell almost half of their power to the state's electric utilities. In the face of falling sugar prices, Hawaiian millers admit that without the revenues from electric power sales, their sugar production would have declined sharply by now. In the continental United States, another 80 megawatts' worth of bagasse-fired plants are operating in Florida and Louisiana.[49]

Researchers at Princeton University estimate that globally some 50,000 megawatts of gas-turbine cogeneration units could be supported with the 1985 level of sugarcane production. South America alone has the potential for 18,000 megawatts. These projections are based on improved gas turbine technologies, the result of recent advances in jet engine designs. In the more than 70 developing countries that grow sugarcane, commercial use of gas turbines could provide as much electricity as the utilities in these nations now generate with oil. The cost of generating electricity with gas-turbine cogeneration units would be lower than for most central-station alternatives.[50]

In most rice-growing developing countries, rice husks are the most abundant crop residue. Every five tons of rice milled produces one ton of husks with an energy content about the same as a ton of wood. At some mills, the residue is burned to generate mechanical energy and steam, but at many it is simply discarded. Using gasifiers or combustion systems, this by-product can be used to power the mill, run irrigation pumps, electrify rural areas, or provide electricity to the grid.

Husk-fired steam power plants are operating in India, Malaysia, the Philippines, Suriname, Thailand, and the United States. A 10.5-megawatt plant in the state of Punjab will operate year-round and burn 20 tons of rice husks per hour. As the world's second largest rice grower, India produces some 18 million tons of hulls annually, enough to justify investment in 500 megawatts of husk-fired generating capacity. The resulting ash is also a valuable product because it is high in silica. Both the Thai and U.S. facilities export their ash to European steel manufacturers for use as an insulator.[51]

Vast paddy areas in Asia are not yet electrified, and mills there depend on imported diesel engines and fuel. The total number of diesel engines used in small rice mills in developing countries proba-

"Globally, some 50,000 megawatts of gas-turbine cogeneration units could be supported with the 1985 level of sugarcane production."

bly exceeds 200,000, with almost 60,000 in Indonesia alone. New husk gasifiers—the result of research and field testing in China, Indonesia, Mali, the Netherlands, the Philippines, Thailand, the United States, and West Germany—may someday be used widely throughout the many rice-growing countries in Asia. A study conducted at the Institute of Technology in Bandung, Indonesia, calculated that the country could save over $30 million each year by replacing relatively low-priced diesel fuel systems with gasifiers.[52]

Other crop residues that could be used more extensively include coconut shells, cotton stalks and ginning waste, peanut and other nut hulls, fruit pits, coffee and other seed hulls, and various sources of straw and fiber. Some food-processing plants already use these residues to provide on-site energy supplies, but there is vast untapped potential. Care must be taken when assessing the viability of these projects to consider the uses to which these materials would otherwise be put: Robbing the land of vital nutrients is counterproductive, whereas producing valuable energy from a by-product that would otherwise be discarded is eminently sensible.[53]

Two additional sources of biomass include food surpluses and plants grown specifically to produce energy. The potential for fast-growing fuel plantations varies greatly among countries because of different resource endowments and sociopolitical systems. Marginal land used to produce energy may be a boon in one setting, but deprive hungry people of land and food in another. One promising approach involves the cultivation of microalgae in otherwise unproductive inland brackish waters. Researchers at the U.S. Solar Energy Research Institute have developed strains that produce up to 66 percent of their body weight as oil.[54]

The potential energy contribution of existing food surpluses is easier to determine, but difficult to extrapolate because of changing agricultural policies, market prices, and weather patterns. The IEA estimates that converting sugar surpluses in the European Community to ethanol would displace 2 percent of liquid petroleum fuels; in the United States, corn surpluses converted to ethanol could replace 7 percent of the country's gasoline consumption.[55]

Brazil and the United States have the world's two largest biomass-based ethanol programs. In Brazil, sugarcane grown specifically for fuel was converted into 10.5 billion liters of ethanol in 1986, providing about half the country's automotive fuel. Most autos burn a gasoline-ethanol mixture that is 20 percent alcohol, but 29 percent of the nation's 10.6 million cars run on pure ethanol.[56]

The large alcohol-fuels program combined with successful offshore oil exploration has enabled Brazil, the largest Third World debtor, to curb its oil imports dramatically and conserve foreign exchange for investment and debt repayment. The ethanol program has created an estimated 475,000 full-time jobs in agriculture and industry, and indirectly another 100,000 jobs in commerce, services, and government. Mechanical power for cane crushers, electricity for on-site needs, and steam for alcohol distillation are provided by cogeneration units that burn the leftover bagasse.[57]

The United States, in contrast, relied on surplus corn and other grains for 90 percent of the 3 billion liters of ethanol it produced in 1987. More than 7 percent of the "gasoline" sold in the country was actually gasohol, a 1-to-9 blend. In the past, ethanol markets have been bolstered by generous tax advantages at the state and national levels and by regulations mandating the reduction of lead in gasoline. Ethanol can replace lead as an octane enhancer.[58]

Alcohol fuels are now gaining support as an air pollution control measure. More than 60 U.S. cities did not meet federal carbon monoxide and ozone standards by the end of 1987. Colorado is the first state to require motorists in its major cities to use gasohol during the winter when pollution is worst. Officials expect carbon monoxide emissions to be cut by 12 percent. Federal legislation has been introduced that would require gasohol use nationally by 1992.[59]

At today's consumption rates, converting just 10 percent of U.S. cars to run solely on alternative fuels would reduce oil imports by 800,000 barrels per day—comparable to what the country imported from Saudi Arabia, Iran, and Iraq in 1986. With 62 percent of U.S. petroleum consumption (more than the country produces) burned for

"At today's consumption rates, converting just 10 percent of U.S. cars to run solely on alternative fuels would reduce oil imports by 800,000 barrels per day."

transportation, renewably based and domestically produced liquid fuels and improved auto fuel economy could reduce significantly the country's dependence on imported oil.[60]

Low prices for sugar and other food crops, a shortage of foreign exchange, and the desire for increased energy independence have prompted at least a dozen other countries to launch alcohol-fuels programs. Except for Canada and the Philippines, the projects are located in Latin America and Africa. Argentina has the world's third largest program, but is able to produce only one-thirtieth as much ethanol each year as Brazil. In Europe, the primary factor behind increased interest in alcohol fuels is the region's tremendous agricultural surplus. Momentum is building to rechannel export subsidies into alcohol-fuels programs. France appears to be taking the lead, with Italy also carefully studying the possibility.[61]

Power From the Sun

Many techniques exist to collect, concentrate, and convert solar radiation into useful energy. The most basic methods use collectors to absorb relatively low-temperature heat and then transfer it to water or air. Somewhat more complicated are systems that concentrate sunlight to produce higher temperature heat for steam or electricity. A third, more sophisticated use of solar power takes advantage of the photovoltaic effect to convert the sun's rays directly into electricity.

Northern Europeans started experimenting with solar collection devices in the seventeenth century to protect tropical plants brought home by explorers from distant lands. Two hundred years later, the first commercial solar product—a water heater—came on the market in the United States. From the early, bare water tanks placed in the sun, solar collector technology has advanced considerably. Today's collectors come in many shapes and sizes; some have transparent cover plates that trap heat without allowing it to be reradiated and that keep out cooling air currents. The heat from the collector is absorbed by a black surface, then transferred to either water or air for immediate use or conveyed to a storage system. Heat loss is reduced by using insulated materials.

Water heating is currently the most popular application of solar collectors; payback times are a short three to five years. Swimming pools and residences are the largest markets, but commercial and industrial systems are becoming more common. In Cyprus, the world's largest solar energy user per capita, private industry has installed solar water heaters on 90 percent of the houses and on a significant portion of apartment buildings and hotels.[62]

In Israel, over 700,000 households, accounting for 65 percent of all domestic water heating, have simple solar systems that cost less than $500 per family to install. Needing only $25 worth of supplemental electric heating per year, the systems pay for themselves in less than four years. To encourage use of this low-cost technology and reduce dependence on imported petroleum, a 1980 law requires the use of solar water heating in all new residential buildings up to nine stories high. In Japan, including more rudimentary devices produced in the sixties, 4 million solar water heaters are in use. In 1984, the peak year for collector sales, 500,000 units were installed. In the remote northwestern region of Australia, 37 percent of households rely on such systems.[63]

Through the mid-eighties, the world's largest collector market was in the United States, where 85 percent of 1984 sales were for residential applications—mostly in Sun Belt states. Unfortunately, by 1986 lower oil and gas prices and the elimination of residential renewable energy tax credits caused the bottom to fall out of the U.S. market. Sales volume dropped more than 70 percent from its 1984 level, and 28,000 of the estimated 30,000 employees in the industry lost their jobs.[64]

The efficiency of flat-plate collectors has increased by 30 percent since 1977, but further gains, along with better materials, are being sought. Lightweight gels and polymer films for glazings and absorber components could cut collector mass sevenfold, thereby reducing production, distribution, and installation costs. If costs come down far

"In Cyprus, the world's largest solar energy user per capita, private industry has installed solar water heaters on 90 percent of the houses."

enough, collectors to heat and cool air, which now account for only 10 percent of the market, may be more widely used.[65]

The second major category of solar technologies encompasses processes that concentrate the sun's rays to produce higher temperatures. Several approaches have been developed that achieve temperatures ranging from 85 to more than 3,000 degrees Celsius. The simplest systems, called solar ponds, rely on lined cavities filled with water and salt; as salt water is denser than fresh water, the salt water on the bottom absorbs heat while water on the surface traps the underlying layer. The most complex systems incorporate thousands of concentrating reflectors that track the sun and focus its light onto a central receiver. A fluid passing through that receiver is used directly to produce heat or indirectly to turn a turbine and generate electricity.

Experimentation and some commercial development of these technologies is taking place in dozens of countries. Israel built the first small electricity generating solar pond in 1979 on the shores of the Dead Sea and in 1984 completed a larger, 5-megawatt pond. An Australian and an American company have each built smaller systems that came on-line in 1985 and 1986 respectively. Ormat Turbines, an Israeli company, is involved in a project to build the world's largest solar pond, in southern California. The unit's 48-megawatt capacity would supply some 40,000 households with electricity.[66]

Other solar thermal energy technologies require significantly more equipment than ponds do. Each type depends on a differently shaped solar reflector to concentrate sunshine. Trough systems use U-shaped mirrors to concentrate the sun's rays onto water- or oil-filled tubes. Most track the sun on one axis and produce temperatures from 100 to 400 degrees Celsius. Concave parabolic dishes follow the sun on two axes and focus the sunlight onto a single point; they yield temperatures as high as 1,700 degrees Celsius.

Central receiver systems use computer-controlled concentrators called heliostats to focus sunlight onto a tower. More than a half-dozen central receiver projects at least 1 megawatt in size have been built with government assistance. (See Table 4.) Though not yet eco-

Table 4: Large Solar Thermal Electric Systems, Constructed or Planned, 1987

Project	Location	Technology	Capacity	Expected or Actual Completion Date
			(megawatts)	(year)
Danby Lake[1]	California	Solar pond	48	—
Luz, SEGS 1	California	Trough Collector	14	1984
Luz, SEGS 2	California	Trough Collector	30	1985
Luz, SEGS 3	California	Trough Collector	30	1986
Luz, SEGS 4	California	Trough Collector	30	1986
Luz, SEGS 5	California	Trough Collector	30	1987
Luz, SEGS 6	California	Trough Collector	30	1988
Luz, SEGS 7–19	California	Trough Collector	450	1989–92
Luz, Eliat	Israel	Trough Collector	25	1990
Solar One	California	Central Receiver	10	1982
Mysovoye	Soviet Union	Central Receiver	5	1986
Bet Ha'Arava	Israel	Solar Pond	5	1984
Solarplant 1	California	Dish Receiver	4	1984
Themis	France	Central Receiver	2	1983
CESA-1	Spain	Central Receiver	1	1983
Sunshine 1	Japan	Central Receiver	1	1981
Sunshine 2	Japan	Hybrid	1	1981
Eurelios	Italy	Central Receiver	1	1981
Solntsye	Soviet Union	Central Receiver	1	1983

[1]Project on hold with no definite completion date.

Source: Worldwatch Institute, based upon research reports, news articles, and private communications.

nomically competitive, studies conducted by several U.S. utilities indicate that costs may come down enough to justify commercial investment around the turn of the century. Even so, this technology is only likely to be economical for large-scale applications.[67]

Dish and trough systems, because of their smaller economies of scale, have been built for less money per unit of generating capacity. A trough system developed by Luz Engineering is enjoying the greatest commercial success. Five of its privately financed Solar Energy Generating Systems already operate in California's Mohave Desert and 14 more are scheduled to come on-line by 1992.[68]

The 30-megawatt plants supply enough electricity for some 10,000 U.S. homes, take only a year to construct, and with supplemental natural gas can consistently provide power during the peak afternoon hours, when utilities value it most. Capital costs have been halved since 1984 and are now comparable to those of recently built U.S. nuclear facilities. The Israeli government has signed a power purchase agreement for a 25-megawatt unit that will be built alongside other solar technologies at a demonstration test site. Serious negotiations are also under way to build a plant in the Indian state of Punjab.[69]

The third major category of solar technologies relies on the photovoltaic effect, discovered by Edmund Becquerel in 1839. This phenomenon causes electricity to be produced when light strikes certain materials. No heat, water, or moving parts are required, just a photon to jar an electron from its orbit, causing an electric current to flow. A 10-percent efficient photovoltaic (PV) cell about 100 square centimeters in size can produce 1 watt of electricity at noon on a clear day. First used to power spacecraft, the terrestrial market now dominates and has grown at an average annual rate of 44 percent from 1980 to 1985.[70]

In 1976, the average market price for a PV module was $44 per peak watt, and a half-million watts' (0.5 megawatts') worth were sold. Only a decade later, costs were down eightfold (in constant 1986 dollars) to $5.25 and shipments had climbed to 24.7 megawatts.[71] (See Figure 1.)

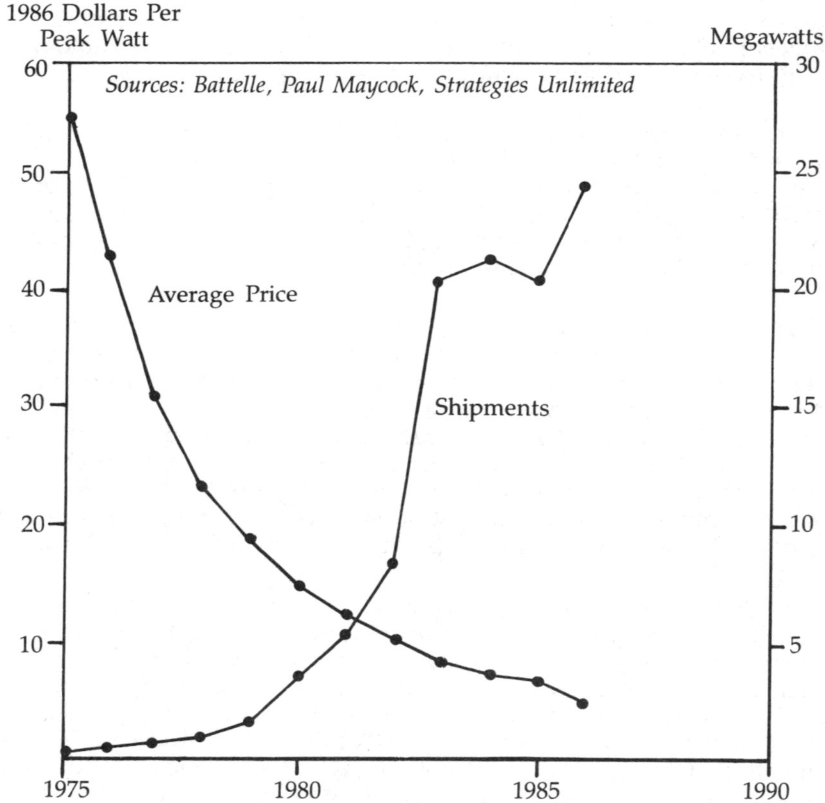

Figure 1: World Photovoltaic Shipments and Average Market Prices, 1975–86

Half the photovoltaics sold in 1986 provide power to equipment or villages not hooked up to electric utility distribution lines. The communications industry accounts for the largest segment of this stand-

"More than 200 million solar-powered calculators were sold in 1987."

alone market and has come to depend on PV systems to reliably relay radio, telephone, and television signals. Multinational communications companies, isolated villages, and researchers in the field all depend on this vital information link. Extending utility lines to remote communication sites can cost between $23,000 and $46,000 per kilometer.[72]

Rural electrification projects that rely on photovoltaics are slowly spreading throughout the Third World. The greatest progress has occurred in the Dominican Republic and on islands in French Polynesia and Greece. Photovoltaics are used to provide refrigeration for life-protecting vaccines, pumping for irrigation projects, and lighting and entertainment for homes. Worldwide, more than 15,000 homes receive their electricity from PV cells, and in West Germany a house was recently built with modules integrated into the roof tiles.[73]

The U.S. Coast Guard owns the largest number of stand-alone PV systems—10,525 used to power navigational aids. Each system saves taxpayers an estimated $5,000 because it eliminates the need to replace batteries at distant lighthouses and ocean buoys. The agency is seeking funding for another 6,400 units in 1988. In some countries, local gasoline stations offer solar-powered battery chargers. And in the Persian Gulf, offshore oil drilling platforms are increasingly powered by photovoltaics. From fence charging to cargo weighing to security systems, PV uses appear limited only by imagination and the cost of alternative energy sources.[74]

The second largest, and newest, category of PV use is for consumer products. Sanyo, Sharp, and Fuji pioneered in this market, and in 1978 the world's first solar-powered calculator was sold. More than 200 million solar calculators were sold in 1987. Though still relatively novel, solar toys, clocks, and backyard lighting systems are all available commercially. These developments were made possible by the discovery of a radically new way of making photovoltaic cells.[75]

Until the eighties, virtually all PV modules were made using some form of single crystal silicon. Although this is the second most abundant element in the earth's crust, the meticulous processing and high degree of purity required—from one part per billion up to less than one part per million of impurities are tolerable—meant that reducing PV costs posed formidable problems.

The major new approach was to develop "thin-film" cells in which the photovoltaic materials are less than one one-hundredth as thick (one micron) as their crystalline counterparts. Made from materials that can be deposited from a gaseous medium onto a substrate, they can be fabricated using continuous production processes. They can also be made in large modules, one foot by three feet. Amorphous silicon modules are presently the most advanced, but research is continuing on many other interesting materials. Lower cost thin-film devices, although not as efficient as single crystal silicon cells, are much cheaper to manufacture and now account for one-third of the total PV market. Japan, which has focused its PV efforts primarily on thin-film consumer products, has seen its market share grow from 15 percent in 1980 to 46 percent in 1986. (See Figure 2.)[76]

The United States produces thin-film cells primarily for power production. ARCO Solar, until 1986 the world's leading PV manufacturer, introduced the first thin-film power module in 1984. The company also has a few consumer products available. Japan only recently branched out into the power market, with systems designed for rural applications in developing countries. The United States, for the moment at least, retains its lead in sales of power generating equipment and provides most of the PV modules used by electric utilities and grid-connected customers—12 percent of the 1986 market.[77]

Together, Japan and the United States account for three-quarters of the world's PV shipments. Brazil, France, India, Italy, the United Kingdom, and West Germany account for most of the rest. Chronar Corporation, a U.S. company, encourages joint venture manufacturing using its equipment. The company has at least partial ownership of plants in China, France, the United Kingdom, and Yugoslavia. One of its first deals was with Alabama Power to build a 1-megawatt PV manufacturing plant in Birmingham. The transaction represented the first direct investment in PV manufacturing made by an American utility.[78]

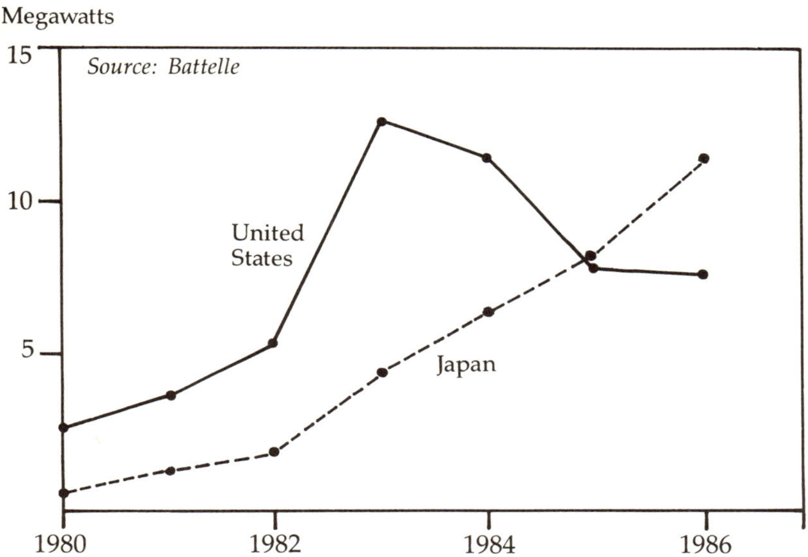

Figure 2: U.S. and Japanese Photovoltaic Sales, 1980–86

Photovoltaic sales will continue to grow at double-digit rates, but a dramatic and widespread increase in their use depends on continued progress in reducing costs, raising efficiencies, improving automated manufacturing techniques, and resolving degradation problems. Commercial PV modules range in efficiency from 7 percent for those made of amorphous silicon to 13 percent for those made of crystalline silicon. By stacking amorphous silicon cells on top of one another, researchers have succeeded in converting 14 percent of the sun's rays into useful power in small solar cells. Theoretically, this level can be doubled. Using concentrated sunlight and crystalline silicon, laboratory cells have achieved efficiencies approaching 28 percent.[79]

The major barrier to wider use of photovoltaics is their cost. At a June 1987 international meeting in California, experts concluded that at today's market prices of $4–5 per watt, the annual worldwide market for photovoltaics was some $125 million. (Adding in the cost of other system components probably raises the 1986 sales figure to approximately twice that much.) These analysts speculated that when the price drops to $3, the market would go to $1.5 billion, and that when it dips below $1 per watt, sales would soar to about $100 billion annually.[80]

ARCO Solar is sufficiently convinced of the industry's growth potential that it is increasing the annual manufacturing capability at one of its plants to 5 megawatts. And Chronar Corporation recently announced plans to build a 10-megawatt production facility. When completed in late 1988, it will be the largest amorphous silicon PV manufacturing plant in the world. This one factory will be capable of producing 40 percent as many PV cells as were traded internationally in 1986.[81]

Tapping the Wind

Early in this century Danish industry relied on wind power for one-quarter of its energy, and 150–200 megawatts of wind capacity were installed throughout the country. Cheap oil and gas and rural electrification displaced these early turbines, and wind energy did not make a resurgence until its rapid growth in the eighties. Electricity producing wind turbines (as opposed to windmills used for mechanical energy) are now installed in 95 countries from the tropics to the Arctic.[82]

Winds are created by the unequal heating of the earth's surface and atmosphere and by the regional differences in pressures that result. Local wind patterns are further influenced by the terrain. Most coasts on every continent and many inland areas are rich in wind resources that, if developed, could supply a major share of the world's electricity.

During the decade following the 1973 oil embargo, well over 10,000 wind machines were installed worldwide. Many of these units are

"At times during the summer of 1987, wind power accounted for more than 5 percent of the power sold by Pacific Gas and Electric."

small and used either to charge batteries or to produce minuscule amounts of electricity, usually less than 100 watts. The market in China for turbines in this size range has grown dramatically, from 1,282 units in 1982 to almost 11,000 in 1986. The demand was created largely by the recent availability of television reception nationwide while much of China has no power grid. Five of the world's 10 largest manufacturers of small wind turbines in 1986 were Chinese.[83]

The international wind power generating market also expanded rapidly in the eighties, from 34 megawatts shipped in 1981 to 567 megawatts in 1985—a seventeenfold increase. By 1986, cumulative worldwide turbine sales reached $2.5 billion. The United States led in the installation of intermediate-size (50–750 kilowatt) wind turbines, and between 1981 and 1986 over 90 percent of the wind energy produced worldwide was sold to the customers of two electric utilities in California.[84]

From 144 machines with a combined capacity of 7 megawatts in 1981, California had 16,769 turbines with a generating capacity of 1,463 megawatts by the end of 1987. (See Table 5.) These turbines were installed in significantly less time than required for traditional generating facilities, and, for the most recent machines, at a lower cost. In 1986, wind turbines generated some 1 percent of the electricity used in northern California.[85]

Most of the turbines installed in the United States are in three mountain passes in California—Altamont, San Gorgonio, and Tehachapi. In Altamont and San Gorgonio, contrary to most other areas, wind speeds are highest in the summer, which coincides with the peak seasonal demand placed on California utilities. At times during the summer of 1987, wind power accounted for more than 5 percent of the power sold by Pacific Gas and Electric, one of the nation's largest electric utilities. According to a recent assessment conducted by the company, less than half the easily developable wind resources in its service territory have been exploited.[86]

Seasonal variations and maximum wind speeds are important in assessing an area's wind power potential because the available power

Table 5: California Wind Farms, 1981–87

Year	Machines Installed	Capacity Installed	Average Capacity	Average Cost	Power Generated[1]
	(number)	(megawatts)	(kilowatts)	(dollars per kilowatt)	(million kilowatt-hours)
1981	144	7	49	3,100	1
1982	1,145	64	56	2,175	6
1983	2,493	172	69	1,900	49
1984	4,687	366	78	1,860	195
1985	3,922	398	101	1,887	670
1986	2,878	276	96	1,250[2]	1,218
1987[2]	1,500	180	120	n.a.	1,600
Total	16,769	1,463	87		3,739

[1]Most wind machines are installed in the last half of a given year and do not produce substantial power until the next year.
[2]Preliminary.

Sources: Pre-1985 from Paul Gipe, "An Overview of the U.S. Wind Industry," *Alternative Sources of Energy*, September/October 1985, based primarily on California Energy Commission (CEC) data; 1985 and most 1986 data from Sam Rashkin, CEC, Sacramento, Calif., private communication, October 6, 1987; preliminary estimates from Paul Gipe, American Wind Energy Association, Tehachapi, Calif., private communication, November 5, 1987.

depends in part on the average wind speed. When it doubles, power output increases eightfold. Most systems are designed to operate in winds of 4–30 meters per second. Potential power output is also proportional to the area swept by each revolution of the turbine's rotor: Doubling the area doubles the power output. Improved turbine design permitted the average rotor diameter of California's machines to reach 17 meters in 1984, resulting in a 50-percent increase in potential power output compared with the 1982 turbine models. Larger rotors effectively lower the capital costs of wind turbines because more electricity can be produced with fewer machines.[87]

During the early eighties many inexperienced manufacturers rushed to cash in on the heavily subsidized California market and built inadequately tested machines. Multiple repairs and major system overhauls soon bankrupted many of these firms. Those that remain have combined sturdy, reliable, proven windmill designs with new materials, knowledge, and manufacturing techniques to create more reliable products. Today's turbines operate 80–98 percent of the time that the wind is blowing.[88]

The average size of turbines installed in California increased from 49 kilowatts in 1981 to 120 kilowatts in 1987. Many of the newest models have capacities of 150–750 kilowatts. Government-funded programs, however, started by promoting large, multimegawatt turbines. Eleven wind machines rated at 1,000 kilowatts or higher had been built as of 1985—seven in the United States, two in Sweden, and one each in Denmark and West Germany. Despite the dismal operating record of some large units, Canada, Denmark, the Netherlands, Sweden, the United Kingdom, and West Germany are building new multimegawatt machines. These governments, partly because of space constraints, are still advocating the use of large turbines, making them the centerpiece of research and development activity.[89]

The reverberations of the U.S. wind energy tax credit termination at the end of 1985 were felt by turbine suppliers worldwide. Outside the United States, Denmark was the hardest hit, as it supplies over half of California's turbines. In 1985, Danish manufacturers shipped more than 2,500 wind units to North America. Wind turbine exports, one of the country's most valuable, fell by half in 1986. U.S. suppliers also suffered setbacks and few U.S. manufacturers remain in business.[90]

The international wind turbine market peaked in 1985 and sales figures are unlikely to return to that level until the nineties. But despite the decline in the California market, interest in wind energy is growing rapidly in other areas of the world. According to a recent international survey, the North American market is expected to gradually decrease in importance to about 50 percent of the global total by the early nineties. Europe will then account for 25 percent or more of the world market, absorbing at least 100 megawatts annually, and the remaining quarter will be geographically dispersed.[91]

Denmark, besides being the leading international supplier of intermediate-size wind turbines, with 7 of the world's top 10 manufacturers, has also built a domestic market. By mid-1987 installed capacity totaled 100 megawatts, including the first sea-based wind plant, on the east coast of Jutland. Although early wind development in Denmark relied almost exclusively on individual 55-kilowatt farm units connected to the grid, the plan for the future is to build clusters of turbines in which each unit will be larger than 200 kilowatts. Elkraft and Elsam, the country's two utility power pools, expect to install 100 megawatts' worth of turbines by 1991.[92]

The Chinese government is calling for wind farms with a total capacity of at least 100 megawatts to be built between 1990 and 1996. In the Netherlands, a five-year plan with the goal of installing 150 megawatts of capacity by 1992 is under way. By the end of the century, the government hopes to have 1,000 megawatts of capacity in place. Spain plans 45 megawatts by 1993 and Greece expects to install 80 megawatts' worth on islands. Smaller wind farms are also either installed or planned in Australia, Belgium, Israel, Italy, the Soviet Union, the United Kingdom, and West Germany.[93]

By far the most ambitious wind energy program belongs to the Ministry of Energy in India, which is pushing to have 5,000 megawatts of wind power generating capacity installed by both public and private developers by the year 2000. The country, which had virtually no wind turbines until 1985, now boasts as much installed capacity as California had in 1981 and is expected to be one of the most rapidly growing markets in the world. If the government achieves its goal, wind may supply more electricity by the end of the century than the country's somewhat optimistic nuclear program.[94]

Average costs for installed intermediate-size wind turbines have fallen by more than half since 1981, to some $800–1,200 per kilowatt. In many markets, such turbines are now competitive with traditional generating technologies. Costs are likely to be reduced further as

"India, which had virtually no wind turbines until 1985, now boasts as much installed capacity as California had in 1981."

more manufacturers, including newly interested Japanese firms, start to mass-produce turbines. Efforts on technical improvements include the development of better blade and shaft materials, and variable-speed constant frequency generators. Some residents near wind farm developments are also demanding a reduction in low-frequency noise.[95]

Laying the Institutional Groundwork

Success at this stage in the development of renewably based energy technologies cannot be judged simply on the share of energy provided or the number of megawatts installed. Technical improvements, including higher efficiencies, increased reliability, and the development of new designs and materials, measure potential contribution. For many of the new technologies, monitoring cost reductions also indicates how much work remains before they achieve commercial acceptance.

Development of new technologies requires sustained research to formulate and test new theories, followed by experimentation with system components and by demonstration projects to assess whether the system is ready for commercial marketing. Because of the risk involved, governments typically pay a large share of these costs, with private contributions increasing as the technology nears commercialization.

The commitment that governments make to R&D on renewables can be measured in several ways. Looking strictly at the dollar figure gives an indication of the amount spent on technical advances. In the United States, still the leader in this category, research and development funding for renewables peaked in 1980 at $900 million (1986 dollars). Appropriations have since fallen 80 percent, with no rebound in sight. Total funding on renewable energy throughout IEA member countries has fallen 64 percent since its 1980 peak. In 1986, R&D funding for renewables fell below 1977 levels for the first time. Greece and Portugal are the only countries whose budgets have continued to increase. In Japan, Switzerland, and Turkey, the drop has been 25 percent or less; in Sweden, West Germany, and the United Kingdom, less than half.[96]

The share of total energy R&D funds going to renewables indicates their relative political support and perhaps the share of new energy supplies that they will contribute. They have strong political backing in Greece, for example, which devotes 63 percent of its energy R&D budget to renewables and does not have a nuclear program. (See Table 6.) Japan, which does have a large nuclear program, devotes only 4 percent of its energy R&D budget to renewables, but nonethe-

Table 6: Government R&D Spending on Renewables in Selected Countries, 1986

Country	Renewables R&D Spending	Share of Energy R&D Budget	Spending Per Capita
	(million dollars)	(percent)	(dollars)
Sweden	17.3	21.8	2.06
Switzerland	10.2	14.7	1.57
Netherlands	17.0	10.6	1.17
West Germany	65.9	11.6	1.09
Greece	9.7	63.2	.97
Japan	99.2	4.3	.82
United States	177.2	7.8	.73
Italy	29.5	3.9	.52
Denmark	2.6	17.8	.51
Spain	19.4	27.6	.50
United Kingdom	16.6	4.4	.29

Sources: International Energy Agency, *Energy Policies and Programmes of IEA Countries: 1986 Review* (Paris: Organisation for Economic Co-operation and Development, 1987); Population Reference Bureau, *1986 World Population Data Sheet* (Washington, D.C.: 1986).

"The learning process requires that new knowledge be accompanied by changes in R&D programs, but sudden cuts in projects and staff can be devastating."

less outspends Greece almost 10 to 1. Japan has 12 times as many people and an economy 17 times larger than Greece's.[97]

Perhaps the best measure for comparing overall government commitment is per capita expenditures. This factors out population size and funding support for other technologies. Sweden, which leads in this category, focuses its efforts on biomass—wood, energy crops, and agricultural surpluses. Because of a national referendum, Sweden must phase out all 12 of its nuclear plants by 2010.[98]

Following the nuclear accident at Chernobyl, government support for renewable energy technologies—particularly those that rely on solar power—has apparently increased in Italy, Japan, Spain, and West Germany. The Swiss government's solar budget is expected to triple by 1990. Denmark and the Netherlands are pushing wind turbines, and Greece and India are launching broad-based renewable energy programs that will rely on a variety of sources.[99]

Widely fluctuating research and development support, as experienced during the eighties, thwarts progress on new technologies. Without reliable funding, the needed long-term programs are difficult to plan and more complicated to implement. That is not to say that there should be no midcourse corrections. The learning process requires that new knowledge be accompanied by changes in R&D programs, but sudden cuts in ongoing research projects and staff can be devastating.

Decisions on how to spend the money that is available are crucial. To open up the research process and stem the preference for established companies and technologies, independent engineering boards could be given the responsibility for selecting the most promising projects.

In Japan, the government and private sector work together closely in developing new technologies. Each contributes monetarily, the government often eases regulatory and fiscal restraints, and the private sector regularly reports—with one voice—on its progress and needs. The New Energy Foundation is a nonprofit organization that includes among its members virtually all the major Japanese manufacturers of photovoltaics. Besides assessing industry progress and pointing out obstacles to growth, it estimates the investments in plant and equipment that will be needed to achieve cost goals over 10- and 20-year periods. Its recommendations concerning research priorities carry significant weight with the government.[100]

One renewable source that has made remarkable strides in recent years because of R&D support is geothermal energy. The use of technologies powered by the heat of the earth's core is growing rapidly in many regions. Of the roughly 5,000 megawatts of capacity installed worldwide, the United States has almost half, followed by the Philippines, Mexico, Italy, and Japan. More than 2,000 additional megawatts are planned to enter service by 1991, and a recent report by the Electric Power Research Institute found that North American capacity alone could reach between 4,200 and 18,700 megawatts by the year 2000. New extraction techniques promise to expand the geographical range in which geothermal energy can be used.[101]

Some of the relatively young renewable energy technologies that have made substantial progress in recent years, but have not achieved the overoptimistic goals set by many in the seventies, include advanced solar collectors, wind turbines, and photovoltaics. Solar collectors and wind turbines are now economical in many settings, but institutional barriers and a financial community whose perceptions lag several years behind technical advances are slowing their penetration of energy markets. Photovoltaic technologies are now competitive in many regions not served by utility power lines. In future decades, they may become standard equipment on new buildings, using the sunlight streaming through windows to generate electricity.[102]

Technologies with long-term potential that require additional, sustained research to demonstrate their feasibility and reduce their costs include biochemical methods of converting biomass feedstocks, various solar thermal configurations, and systems that make use of the temperature difference between surface and deep ocean waters. Development of an economical storage method that could stockpile electricity for later use would also be a boon for renewables. Wind and sunlight are not always available when needed, but if their energy could be stored for later use, their intermittent nature would pose less of a problem.

Two possible storage methods include "pumped hydro" and chemical storage in batteries. Pumping water uphill during periods of low electricity demand in order to generate power during peak demand is a common practice in the United States, where more than 16,000 megawatts' worth of such facilities have been installed. The practice is starting to be adopted in other countries, but thus far few have installed significant pumped hydro capacity. Battery storage is a newer technology just now being tested. Southern California Edison started construction of the world's largest lead-acid battery storage system in August 1987. The battery will be able to supply 10 megawatts of power for four hours, enough to meet the needs of 5,000 customers. The system is projected to cost $13.5 million. Future plants using advanced batteries made with different materials are expected to cost only half as much.[103]

Developing new technologies is a requisite step to ensure their use, but not a sufficient one. In most countries, the electric power industry is composed of either government-owned or government-regulated monopolies badly in need of reform. Averse to change, these institutions are slow to adopt new technologies, policies, and pricing systems. Electricity prices do not reflect market value; rather, they are determined by fiat or by regulation, and often bear little relation to the cost of building new power plants. Utilities have nonetheless proved resilient, and it is only in the United States that fundamental change has begun.

The Public Utility Regulatory Policies Act (PURPA), passed by the U.S. Congress in 1978, made it possible for a wide range of newcomers to compete in the power industry. Utilities were required to purchase power made available by independent producers at the "avoided cost" of utility generation. In California, where the act was most vigorously enforced, private developers aided by tax credits not only found it economical to build virtually all of the state's new generating capacity, they are now displacing existing power plants.

By the fall of 1987, private developers of renewable energy projects had requested regulatory approval to build 16,335 megawatts of capacity. By the end of 1985, over 6,000 megawatts' worth had already been built. Biomass made the greatest contribution, followed by wind and small hydropower. When combined with technology installed by the nation's utilities, 14 percent of the 1985 electricity generating capacity in the United States relied on renewable energy sources. (See Table 7.) Including nonelectric applications, renewables contribute about as much total energy in the United States as Sweden and Portugal together derive from all energy sources.[104]

Unfortunately, with the cessation of most renewable energy tax credits and with low natural gas prices, the independent power industry is starting to turn away from renewably based projects in favor of natural gas cogeneration systems. The share of proposed new generating capacity that would rely on renewables is falling steeply—down from 29 percent in 1986 to 12 percent in 1987—and changes are being considered that would increase the competition among large and small power plants and probably further reduce the role of renewables.[105]

One of PURPA's goals was to diversify the nation's energy supply and encourage the use of renewable energy sources. To ensure that the private power industry does not focus all its efforts on fossil-fuel-based projects, with the lowest initial costs, some incentives or new regulatory initiatives are required that would effectively reserve a share of new generating capacity for renewables. Innovation itself has value, as do technologies such as wind and geothermal power that do not use fuel, do not pollute, and are invulnerable to oil price increases.

The competitive power industry in the United States has already encouraged innovation. Independent production of electricity has spurred private investment by providing the first commercial market for renewable-energy-based generating plants. Not only are costs falling, but reliability is improving in facilities that are on average 50 times smaller than the fossil- and nuclear-fuel-based generating plants formerly built by electric utilities. Smaller increments of gener-

Table 7: United States: Renewably Based Electricity Generating Capacity, 1985

Fuel Type	Independently Produced	Utility Produced	Total
		(megawatts)	
Wood	1,674	200	1,874
Wind	1,283	24	1,307
Small Hydropower	943	2,247	3,190
Waste[1]	935	150	1,085
Geothermal	226	1,580	1,806
Solar	82	13	95
Other	871	—	871
Large Hydropower	—	86,616[2]	86,616
Total Renewables Capacity	6,014	90,830	96,844
Total Generating Capacity	20,062	656,118	676,180
		(percent)	
Share of Capacity That Is Renewably Based	30	14	14

[1]Includes municipal solid waste and agricultural residues.
[2]Includes 16,356 megawatts of pumped storage capacity.

Sources: Edison Electric Institute, "1985 Capacity and Generation Non-Utility Sources of Energy," Washington, D.C., April 1987; Douglas Cogan and Susan Williams, *Generating Energy Alternatives, 1987 Edition* (Washington, D.C.: Investor Responsibility Research Center, 1987); U.S. Department of Energy, *Inventory of Power Plants in the United States 1985* (Washington, D.C.: 1986); pumped storage capacity of large hydropower from Jean Jarecki, U.S. Federal Energy Regulatory Commission, Washington, D.C., private communication, September 30, 1987.

ating capacity, which can be built in considerably less time than large, conventional facilities, lessen both the risk and difficulty of accommodating unpredictable growth in power demand. A 1987 study by the California Energy Commission found that many of these technologies are already more economical than conventional nuclear and coal plants.[106]

Legislation designed after PURPA, though less comprehensive, is now on the books in Greece and New Zealand. Proponents hope to stimulate wind and solar projects. Denmark, India, Israel, the Netherlands, Pakistan, Turkey, the United Kingdom, and West Germany are promulgating new rules of their own, and legislation has been proposed in Sweden. In Nepal and China private entrepreneurs are encouraged to build and operate small-hydropower facilities, both to sell power to the grid and to introduce it in nonelectrified areas.[107]

Increasing the Contribution from Renewables

Only periodically throughout history has reliance on renewable energy sources been interrupted by the discovery of apparently plentiful and cheap fuels, such as new deposits of coal, oil, natural gas, and uranium. But coal burning is warming the atmosphere and producing acid rain that is implicated in damaging 31 million hectares of forests in central Europe. Nuclear power is producing radioactive waste that will be hazardous for millennia. And oil wells are not replenishable on a human time scale.[108]

Despite dwindling global supplies and the oil price jolts of the seventies, many countries are increasing their dependence on the volatile Middle East. When oil prices resume their upward climb, countries that did not institute policies to strengthen their energy independence may be more vulnerable to supply disruptions than they were in the seventies. The leaders of these nations will be rightly blamed for failing to prepare for widely foreseen chaos in world oil markets.

In addition, the competition for remaining oil reserves will be greater. According to a recent study published by the World Resources Institute, if per capita Third World commercial energy use continues to grow at an annual rate of 3.6 percent, as it did in the seventies, energy use per capita by 2020 would still be only 57 percent of what it was in

"The costs of acid rain damage, climate change, and diminished security warrant higher investments in alternatives to fossil fuels."

Western Europe in 1975. But if in addition, the population of developing countries nearly doubles, as expected, the demand for commercial energy in those countries would increase 7.5 times from 1980 to 2020. The increment in energy use by developing countries would equal 1.3 times the world's total energy use in 1980, three times its oil production, or five times its coal production. The authors point out that "It would be exceedingly difficult to meet such energy requirements at reasonable costs and without major environmental or security problems."[109]

The costs of acid rain damage, climate change, and diminished security warrant higher investments in alternatives to fossil fuels. Increasing the contribution of renewables requires market pricing of all energy sources, greater R&D funding from both the public and private sector, and institutional restructuring to open electricity markets and remove barriers to the use of renewables.

The most effective way to open the market for renewables is to get the price signals right. Such subsidies as artificially low prices for diesel fuel and kerosene, incentives for coal mining and transportation, tax breaks for petroleum producers, and below-cost electricity prices all hamper the development and commercial attractiveness of alternatives.

In China, for instance, consumer subsidies for heavy fuel oil and crude oil equal $5.4 billion, electricity subsidies amount to $8.9 billion, and coal subsidies total $10.4 billion. Total energy subsidies equal $19.4 billion—equivalent to 7 percent of China's gross national product (GNP) and 20 percent of its export earnings. Energy subsidies are even higher in Egypt, where they equal 13 percent of GNP and 88 percent of export revenues. Elsewhere, military involvement to keep open petroleum supply lines and taxpayer-funded insurance policies for accidents at nuclear reactors should be tallied as energy subsidies.[110]

In the absence of competitive markets, governments may sometimes want to explicitly encourage the use of new technologies. This may be done by offering fiscal incentives such as tax credits or by establishing government-funded testing laboratories that help assure consumers of the reliability of new technologies.

In the United States, the availability of renewable energy tax credits for businesses and residences from 1978 through 1985 spurred the markets for wind turbines, small hydro facilities, geothermal projects, solar collectors, and photovoltaic cells. Unfortunately, instead of being gradually diminished in value, most of these tax credits were yanked out wholesale—a devastating disruption to poorly financed renewable energy industries. Denmark, which offers similar credits to wind project developers, is phasing them out incrementally to ease the transitional shock to the industry. The standard setting approach has also been adopted. Israel, Spain, and the United States, for example, test solar collector technologies, and Denmark and the Netherlands have certification programs for wind turbines. In 1985, the Danish government provided the wind industry with $6 million in total subsidies. Wind turbine sales of about $250 million were some 42 times greater in value.[111]

Many renewable energy technologies already make a substantial contribution to global energy supplies. Biomass fuels alone now supply 15 percent of the world's energy. In developing countries their share triples, though often the supplies are not managed renewably and the fuel is burned wastefully. Like all energy sources, renewables will make the greatest contribution when used efficiently. Anaerobically fermenting animal dung, for example, provides both biogas for residential cooking and lighting and a better fertilizer than raw dung. New biomass gasifiers that produce both heat and electricity produce more useful energy per unit of fuel than traditional combustion systems.[112]

One country that has made increased reliance on renewables a government priority is the Philippines, where half the energy and 60 percent of the electricity used in 1984 were supplied by biomass, hydropower, and geothermal energy. Oil price hikes in the seventies prompted the government to actively seek alternatives to the country's heavy dependence on imported petroleum. Renewable sources of energy accounted for 87 percent of the net increase in energy

"In the Philippines, renewable sources accounted for 87 percent of the net increase in energy supply between 1974 and 1984."

supply between 1974 and 1984. Programs were launched to burn and gasify rice hulls, sugarcane residues, wood, and coconut husks and shells. The Alcogas program to produce ethanol from sugarcane has had a rocky history but received renewed support from the Aquino government in October 1986. The strategy aims not only to reduce petroleum imports but also to aid the depressed sugar industry and to lower the average levels of lead in the blood of Filipinos.[113]

The greatest government support in the Philippines has gone to geothermal energy. Installed capacity grew from 3 megawatts in 1978 to 894 megawatts in 1986. Another 1,300 megawatts of capacity are under construction or planned. The government has worked closely with private industry to develop this energy source; it conducts all survey and exploration work, exempts developers from duties and taxes on imported and purchased equipment, and acts as a cofunder and profit-sharer on commercial projects.[114]

In India, slightly less than half of total energy use is provided by renewables, but most of that is noncommercial fuelwood, agricultural wastes, and animal dung. The country's Department of Nonconventional Energy Sources has drawn up a plan to build 15,000 megawatts of renewably based electricity capacity by the year 2000. (This is in addition to the country's large hydro program.) Wind turbines and biomass technologies would make the greatest contributions, followed by small hydroelectric and solar thermal power stations and an even smaller share from biogas and municipal refuse burning facilities. By the turn of the century, small-scale renewable energy technologies will contribute more to India's commercial energy supplies than will nuclear power.[115]

Unfortunately, because of the international debt crisis, few developing countries are perceived as good prospects for commercial lending. With investments in traditional enterprises dramatically slowed, convincing bankers of the merits of renewable energy systems will be difficult at best. But if international lending agencies, such as the World Bank, were to get behind investments in small-scale, locally appropriate, renewable energy technologies, a proposition that is under consideration, private financial institutions might be persuaded to join in.

The purpose of the United Nations Conference on New and Renewable Sources of Energy in 1981 was to encourage widespread adoption of renewable energy sources in developing countries, but the follow-up program fell victim to domestic political agendas and bureaucratic inertia. Although some bilateral and multilateral aid programs have effectively promoted the development of indigenous energy resources, their efforts need to be strengthened and expanded if real change is to occur.

During the next decade, the contribution of renewable energy sources is expected to increase significantly in a handful of countries. (See Table 8.) Some of the relative newcomers to this group include Denmark, Greece, Israel, Japan, and West Germany. In many of these nations, the government has made increased use of renewables a priority and has backed up the commitment with funding and policy support.[116]

Too many countries, however, still have no vision of a sustainable energy future. Long-term planning ranges from solving daily crises to drawing up next year's budget. Lack of foresight results in the security risk of unreliable supply, the possibility of military engagement to protect energy sources and transportation routes, and inattention to the environmental consequences of using conventional energy sources.

Although the nuclear accident at Chernobyl prompted some governments to seriously reevaluate their energy strategies, the rethinking and economic restructuring has not gone far enough. Most countries still depend on fossil- and nuclear-fired generating equipment, with 20- to 30-year payback periods, that is installed by monopolistic utilities. Policymakers are not taking advantage of the increased economic competitiveness of many renewable energy technologies. The absence of a coherent, long-range energy policy cripples private investment in capital-intensive renewables technologies, despite their minimal operating and fuel costs.

Table 8: Share of Total Primary Energy Requirements Supplied by Renewables in Selected Countries, 1984/85, with Projections for 2000

Country	1984/85	2000
	(percent)	
Brazil[1]	59.0	64.3
Norway	61.1	63.0
Japan	5.1	13.5
Australia	9.4	12.6
Israel	2.3	12.0
Denmark	2.0	10.0
Greece	5.9	8.9
United States	7.4	8.7
West Germany	2.5	5.5

[1]Figures for 1983 and 1993.

Sources: Ministry of Mines and Energy, "Energy Self-Sufficiency: A Scenario Developed as an Extension of the Brazilian Energy Model," Government of Brazil, Brasilia, 1984; Strategies Unlimited, "International Energy and Trade Policies of California's Export Competitors," California Energy Commission, Sacramento, Calif., 1987; Scott Sklar, "International Trade Policy for the Renewable Energy Industries: An Assessment," *Solar Today*, March/April 1987; International Energy Agency, *Energy Policies and Programmes of IEA Countries: 1986 Review* (Paris: Organisation for Economic Co-operation and Development, 1987).

A sustainable energy path that relies on renewables and energy efficiency will provide policymakers the flexibility to cope with an uncertain global future. Those ready to make the change need to improve market pricing signals, open up the energy supply and energy savings business, and reinvigorate research and development programs. Educating their colleagues as well as consumers to see the true present value of future savings, and tailoring financial practices to reflect this vision, will be difficult. Nations that accept the challenge will be rewarded with increased energy security, more stable economies, and a healthier global environment.

Notes

1. Sandra Postel, *Altering the Earth's Chemistry: Assessing the Risks*, Worldwatch Paper 71 (Washington, D.C.: Worldwatch Institute, July 1987); International Co-operative Programme on Assessment and Monitoring of Air Pollution Effects on Forests, "Forest Damage and Air Pollution: Report on the 1986 Forest Damage Survey in Europe," Global Environment Monitoring System, United Nations Environment Programme, Nairobi, mimeographed, 1987; Ralph Rotty, University of New Orleans (formerly of Institute for Energy Analysis, Oak Ridge Associated Universities, Oak Ridge, Tenn.), private communication, June 16, 1987.

2. Worldwatch Institute estimate based on Rotty, private communication; on J.A. Edmonds and J.M. Reilly, "A Long-Term Global Energy-Economic Model of Carbon Dioxide Release from Fossil Fuel Use," *Energy Economics*, April 1983; and on United Nations Department of International Economic and Social Affairs, *1985 Energy Statistics Yearbook* (New York: 1987).

3. Global contribution of renewables from World Commission on Environment and Development, *Our Common Future* (New York: Oxford University Press, 1987); Third World biomass figure from D.O. Hall et al., *Biomass for Energy in Developing Countries* (Elmsford, N.Y.: Pergamon Press, 1982).

4. Investment figure is Worldwatch Institute rough estimate, assuming investments of $20 billion in hydropower, $6 billion in biomass, $2.5 billion in geothermal, $550 million in solar collectors, $500 million in R&D, $400 million in wind power, and $250 million in photovoltaics.

5. Ministry of Mines and Energy, "Energy Self-Sufficiency: A Scenario Developed as an Extension of the Brazilian Energy Model," Government of Brazil, Brasilia, 1984; United Nations, *1985 Energy Statistics*; U.S. Department of Energy (DOE), *International Energy Annual 1986* (Washington, D.C.: 1987); José Goldemberg et al., *Energy for Development* (Washington, D.C.: World Resources Institute, 1987).

6. Goldemberg et al., *Energy for Development*.

7. Third World reliance on imported oil from U.S. Agency for International Development (AID), *Decentralized Hydropower in AID's Development Assistance Program* (Washington, D.C.: 1986).

8. United Nations, *1985 Energy Statistics*.

9. Guri dam information from "Price Decline is Harmful for Development of Energy Resources," *OPEC Bulletin*, May 1987; Peter T. Kilborn, "Brazil's Hy-

droelectric Project," *New York Times*, November 14, 1983; Catherine Caulfield, "The Yangtze Beckons the Yankee Dollar," *New Scientist*, December 5, 1985; Strategies Unlimited, "International Market Evaluations: Small-Scale Hydropower Prospects," California Energy Commission (CEC), Sacramento, 1987.

10. T. W. Mermel, "Major Dams of the World-1986," *Water Power & Dam Construction*, July 1986; World Energy Conference, *Survey of Energy Resources, 1980* (Munich: 1980).

11. "The World's Hydro Resources," *Water Power & Dam Construction*, October 1986; Gary Aderman, "China Turns to Hydropower," *Journal of Commerce*, October 1, 1987; Brazil Ministry of Mines and Energy, "Energy Self-Sufficiency."

12. Marlise Simons, "Dam's Threat to Rain Forest Spurs Quarrels in the Amazon," *New York Times*, September 6, 1987; Catherine Caulfield, "Dam the Amazon, Full Steam Ahead," *Natural History*, July 1983; Philip M. Fearnside, National Institute for Research in the Amazon, Manaus, Brazil, private communication, May 29, 1987; "Rainforest Sacrificed to Build Amazon Dam," *World Wildlife Fund News*, July/August 1984.

13. Kilborn, "Brazil's Hydroelectric Project"; World Bank, *A Survey of the Future Role of Hydroelectric Power in 100 Developing Countries* (Washington, D.C.: 1984); Don Winston, U.S. Council on Energy Awareness, Washington, D.C., private communication, October 7, 1987.

14. Donald Worster, "An End to Ecstasy: What Will the Dam Builders Do Now?" *Wilderness*, Fall 1987; Philip Shabecoff, "U.S. Bureau for Water Projects Shifts Focus to Conservation," *New York Times*, October 2, 1987.

15. "Canadian National Energy Board Rejects H-Q Hydroelectric Sale to New England," *International Solar Energy Intelligence Report*, June 23, 1987; "Canadians Size Up U.S. Hydro Export Market," *Alternative Sources of Energy*, July/August 1987; Bill Rankin, "Manitoba Hydro Plans Large Exports of Electricity to the U.S.," *Energy Daily*, January 8, 1986.

16. "Where Dams Can Cause Wars," *The Economist*, July 18, 1987; "World Status: Middle Eastern Electricity," *Financial Times Energy Economist*, May 1987.

17. Catherine Caulfield, "Environmentalists Warn of Damage from Planned Dam in China," *Christian Science Monitor*, December 9, 1985; Claude Alvares and Ramesh Billorey, "Damming the Narmada: The Politics Behind the Destruction," *The Ecologist*, May/June 1987. For furthur information see Bruce

Rich, Environmental Defense Fund, Testimony in Hearings on Environmental Performance of Multilateral Development Banks, Subcommittee on International Development Institutions and Finance, U.S. House of Representatives, April 8, 1987.

18. Edward Goldsmith and Nicholas Hilyard, *The Social and Environmental Effects of Large Dams* (San Francisco, Calif.: Sierra Club Books, 1987);Philip B. Williams, "Damming the World," *Not Man Apart*, October 1983; Caulfield, "Environmentalists Warn of Damage"; Robert Goodland, *Environmental Assessment of the Tucurui Hydroproject* (Brasilia: Electronorte, 1978).

19. Nicomedes D. Briones and Jose P. Castro, "Effective Management of a Tropical Watershed: The Case of the Angat River Watershed in the Philippines," *Water International*, December 1986; Andrew Maguire and Janet Welsh Brown, eds., *Bordering on Trouble: Resources and Politics in Latin America* (Bethesda, Md.: Adler & Adler, 1986); International Task Force, *Tropical Forests: A Call for Action, Part 1: The Plan* (Washington, D.C.: World Resources Institute, 1985).

20. Tony Davis, "Managing to Keep Rivers Wild," *Technology Review*, May/June 1986; Richard M. Glick, "Decommissioning the Hydro Industry in the Pacific Northwest," *Alternative Sources of Energy*, October 1986; Ted Williams, "The Salmon Be Dammed," *Resources*, October 1983; Alan Long, Summary of Appalachian Mountain Club Testimony in Hearings on Federal Energy Regulatory Commission Small Hydro Program, Subcommittee on Energy Conservation and Power, U.S. House of Representatives, August 30, 1984.

21. Frost & Sullivan cited in Don Best, "Remote Power Market Is Predicted to Swell," *Renewable Energy News*, July 1985.

22. World Bank, *Survey of Hydroelectric Power*; Strategies Unlimited, "Small-Scale Hydropower Prospects"; Larry N. Stoiaken, "The Chinese Hydro Imports: Testing the North American Marketplace," *Alternative Sources of Energy*, July/August 1983.

23. Edison Electric Institute, "1985 Capacity and Generation Non-Utility Sources of Energy," Washington, D.C., April 1987; Douglas Cogan and Susan Williams, *Generating Energy Alternatives, 1987 Edition* (Washington, D.C.: Investor Responsibility Research Center, 1987); Donald Marier and Larry N. Stoiaken, "An Industry in Transition: The Hydropower Industry Looks Ahead," *Alternative Sources of Energy*, July/August 1987; "Poland Restarts Small Hydro Plants," *European Energy Report*, July 24, 1987; Ian Lewis, "Small

58

Hydro Playing Key Role in Ontario's Economy," *Alternative Sources of Energy*, October 1986.

24. AID, *Decentralized Hydropower*; CEC, "Relative Cost of Electricity Production," Staff Report, Sacramento, Calif., April 1987; Strategies Unlimited, "Small-Scale Hydropower Prospects"; Maria Elena Hurtado, "Hydro Power: China's Marriage of Convenience," *South*, January 1983.

25. Worster, "An End to Ecstasy."

26. James Fay, "Harnessing the Tides," *Technology Review*, July 1983; Electric Power Research Institute, "Ocean Energy Technologies: The State of the Art," Palo Alto, Calif., 1986.

27. For information on La Rance plant, Annapolis Royal plant, and Chinese program, see International Energy Agency (IEA), *Renewable Sources of Energy* (Paris: Organisation for Economic Co-operation and Development (OECD), 1987); furthur information on Annapolis Royal from George Hagerman, Sea-Sun Power Systems, Alexandria, Va., private communications, October and November 1987; information on the Soviet program from Mark Newham, "Soviets Could Move Swiftly to Heavy Reliance on Solar," *Solar Energy Intelligence Report*, May 13, 1986.

28. David A. Greenberg, "Modeling Tidal Power," *Scientific American*, November 1987; Fox Butterfield, "Nova Scotia Dam's Effect on Coast Ecology Feared," *New York Times*, August 6, 1984; Hagerman, private communication.

29. Roger Milne, "Testing Ground for the Severn's Tidal Power," *New Scientist*, September 24, 1987; "Tidal Power—Small is Beautiful?" *Network for Alternative Technology and Technology Assessment Newsletter*, March/April 1987.

30. IEA, *Renewable Sources*.

31. David Ross, "World's First Commercial Wavepower Stations Sold," *Alternative Sources of Energy*, May/June 1987.

32. David Ross, "Wave Power," *Network for Alternative Technology and Technology Assessment Newsletter*, July/August 1987; Mark Newham, "In Switch, Britain to Fund Wave Energy Project," *International Solar Energy Intelligence Report*, July 14, 1987.

33. Malcolm W. Browne, "Wave Energy Converted into Electricity," *New York Times*, February 10, 1987; Takeaki Miyazaki, "Wave Power Generator Kaimei," *Oceanus*, Spring 1987; Hagerman, private communication.

34. J. Hilbert Anderson, "Ocean Thermal Power—The Coming Energy Revolution," Sea Solar Power, Inc., York, Pa., unpublished paper, July 1982; Terry R. Penney and Desikan Bharathan, "Power from the Sea," *Scientific American*, January 1987.

35. Penney and Bharathan, "Power from the Sea"; L.L. Lewis et al., "Open-Cycle OTEC Seawater Experiments in Hawaii," presented at Oceans '87, Halifax, Nova Scotia, September 28–October 1, 1987.

36. Hagerman, private communication.

37. IEA, *Renewable Sources;* World Commission on Environment and Development, *Our Common Future.*

38. Gordon T. Goodman, "Biomass Energy in Developing Countries: Problems and Challenges," *Ambio*, Vol. 16, No. 2–3, 1987; United Nations, *1985 Energy Statistics.*

39. IEA, *Renewable Sources.*

40. Goodman, "Biomass Energy"; U.N. Food and Agriculture Organization, *Fuelwood Supplies in the Developing Countries*, Forestry Paper 42 (Rome: 1983).

41. National Wood Energy Association, "Wood Energy—America's Renewable Source," fact sheet, Washington, D.C., September 1987; American Paper Institute, "U.S. Pulp, Paper and Paperboard Industry Estimated Fuel & Energy Use, Full Year 1986, 1985 and 1984," New York, April 1987; Robert P. Kennel, "Biomass for Cogeneration (A Better Option Than You Expected)," presented at Co-energy '86, Boston, Mass., September 3–4, 1986; Solar Energy Industries Association, *Energy Innovation: Development and Status of the Renewable Energy Industries, Vol. 2* (Washington, D.C.: 1985).

42. Meridian Corporation, *Electric Power From Biofuels: Planned and Existing Projects in the United States* (Washington, D.C.: DOE, 1985); Everett Jordan, Eugene Water & Electric, Eugene, Ore., private communication, September 3, 1987; Thomas Carr, Burlington Electric Department, Burlington, Vt., private communication, August 12, 1987; Gerry Anderson, Northern States Power Company, Minneapolis, Minn., private communication, September 2, 1987; George Parks, Washington Water & Power, Kettle Falls, Wash., private communication, September, 1987; CEC, "Relative Cost."

43. Robert P. Kennel, "Comments of the National Wood Energy Association on Cogeneration and Small Power Production," before the Federal Energy Regulatory Commission, April 30, 1987; Kennel, "Biomass for Cogeneration."

44. Kennel, "Comments of National Wood Energy Association"; Robert P. Kennel, Vice President, Ultrasystems Development Corporation, Fairfax, Va., private communication, September 2, 1987.

45. Dean Mahin, "Wood-Fuel Users Report Cost Savings in Virginia," *Renewable Energy News*, October 1985.

46. Frank H. Denton, *Wood for Energy and Rural Development: The Philippines Experience* (Manila: Frank H. Denton, 1983); Christopher Flavin, "Bio-Energy in the Philippines," Worldwatch Institute, unpublished memorandum, December 1985.

47. IEA, *Renewable Sources*.

48. Al Binger, President, Biomass Users Network, Washington, D.C., private communication, October 1, 1987; Bill Belleville, "Renewable Energy Promises Much As Caribbeans Look to the Future," *Renewable Energy News*, October 1985; Eric Larson, Center for Energy and Environmental Studies, Princeton University, Princeton, N.J., private communication, November 12, 1987; information on Thailand from RONCO Consulting Corp., "The Sugar Industry in the Philippines," Arlington, Va., December 1986.

49. Information on Hawaii from Charles Kinoshita, Hawaiian Sugar Planters Association, Honolulu, Hawaii, private communication, November 24, 1987, and from RONCO, "Sugar Industry in the Philippines"; information on continental capacity from Michael D. Devine et al., *Cogeneration and Decentralized Electricity Production* (Boulder, Colo.: Westview Press, 1987).

50. Eric D. Larson et al., "Steam-Injected Gas-Turbine Cogeneration for the Cane Sugar Industry: Optimization Through Improvements in Sugar-Processing Efficiencies," Center for Energy and Environmental Studies, Princeton University, Princeton, N.J., September 1987.

51. AID, "Power from Rice Husks," *Bioenergy Systems Report*, April 1986; "Rice-hull-fired Powerplant Burns a Nuisance Waste, Sells Electricity, Ash," *Power*, July 1985.

52. AID, "Power from Rice Husks."

53. Per Johan Svenningsson, "Cotton Stalks as an Energy Source for Nicragua," *Ambio*, Vol. 14, No. 4–5, 1985; Amory B. Lovins et al., "Energy and Agriculture," in Wes Jackson et al., eds., *Meeting the Expectations of the Land* (San Francisco, Calif.: North Point Press, 1984).

54. H.M. Hubbard, Director, Solar Energy Research Institute, Testimony before the Subcommittee on Energy Research and Development, U.S. House of Representatives, March 11, 1987.

55. IEA, *Renewable Sources.*

56. Dr. Marcos M. Soares, Technical Assistant, National Executive Commission of Alcohol, Government of Brazil, Brasilia, private communication, June 25, 1987; "Fuel Consumption High Despite Price Hike," *Gazeta Mercantil*, January 26, 1987; National Executive Commission of Alcohol, "The National Alcohol Program," Ministry of Industry and Commerce, Government of Brazil, Brasilia, 1984.

57. Robert H. Williams, "Potential Roles for Bioenergy in an Energy Efficient World," *Ambio*, Vol. 14, No. 4–5, 1985; Howard S. Geller, "Ethanol Fuel From Sugar Cane in Brazil," in Annual Reviews Inc., *Annual Review of Energy, Vol. 10* (Palo Alto, Calif.: 1985).

58. Information Resources, Inc., information packet, Washington, D.C., 1987; A. Barry Carr, Congressional Research Service, Testimony at Hearings on Possible Effects of Legislation Mandating Use of Ethanol in Gasoline, Subcommittee on Energy and Power, U.S. House of Representatives, June 24, 1987; Richard B. Schmitt, "Gasohol Backers See Ban on Lead Boosting Sales," *Wall Street Journal*, September 26, 1985; Sarah McKinley, "Ethanol Enjoys Good Times, But Is There A Hangover Ahead?" *Energy Daily*, August 21, 1985.

59. Cities not meeting U.S. pollution standards from Brock Nicholson, U.S. Environmental Protection Agency, Research Triangle Park, N.C., private communication, November 9, 1987; Mark Ivey and Ronald Grover, "Alcohol Fuels Move Off the Back Burner," *Business Week*, June 29, 1987.

60. Jay Rockefeller and Phil Sharp, "Save Oil, Fight Ozone Pollution," *Washington Post*, July 1, 1987; Mary C. Holcomb et al., *Transportation Data Book: Edition Nine* (Oak Ridge, Tenn.: Oak Ridge National Laboratory, 1987).

61. Meridian Corporation, "Worldwide Review of Biomass Based Ethanol Activities," Falls Church, Va., 1985; Information Resources, Inc., *Alcohol Outlook*, various issues; Alan Friedman and George Graham, "Ferruzzi Plans to Produce Ethanol at Plant in Northern France," *Financial Times*, July 10, 1987;

David Lindahl, DOE, Washington, D.C., private communication, August 27, 1987; "Alcoholic Problems in Italy," *European Energy Report*, August 21, 1987.

62. IEA, *Renewable Sources;* Ross Pumfrey and Thomas Hoffman, "Incentives for the Use of Renewable Energy: The Experience in Brazil, Cyprus, India, the Philippines, and California," International Institute for Environment and Development, Washington, D.C., 1985.

63. D. Groues and I. Segal, *Solar Energy in Israel* (Jerusalem: Ministry of Energy & Infrastructure, June 1984); IEA, *Renewable Sources;* Strategies Unlimited, "International Market Evaluations: Solar Thermal Energy Prospects," CEC, Sacramento, Calif., 1987.

64. DOE, *Solar Collector Manufacturing Activity 1986* (Washington, D.C.: 1987); DOE, *Solar Collector Manufacturing Activity 1984* (Washington, D.C.: 1985); Scott Sklar, Solar Energy Industries Association, Testimony at Hearings, Subcommittee on Energy Research and Development, U.S. House of Representatives, March 11, 1987.

65. IEA, *Renewable Sources*.

66. Michael Edesedd, "On Solar Ponds: Salty Fare for the World's Energy Appetite," *Technology Review*, December 1982; "Bet Ha'Arava Solar Pond Power Plant Inaugurated," *Sun World*, Vol. 8, No. 1, 1984; "Solar Ponds Performing Well, Several Countries Advance Plans," *Solar Energy Intelligence Report*, April 28, 1987; Robert L. Reid and Andrew H.P. Swift, "El Paso Solar Pond First in U.S. to Generate Electricity," *Solar Today*, January/February 1987; "California Looks to Salt Water and the Sun," *New Scientist*, July 3, 1986.

67. CEC, "Relative Cost"; "APS, PG&E Settle on Molten Salt for Solar Thermal Central Receiver," *Solar Energy Intelligence Report*, September 8, 1987.

68. Trudy Self, Luz International Limited, Los Angeles, Calif., private communication, July 23, 1987; "Cogeneration/Small Power Projects, Quarterly Report," Southern California Edison, Rosemead, Calif., June 30, 1987.

69. Luz International Limited, information packet, Los Angeles, Calif.; Self, private communication; "Solar Energy Strikes Gold in California," *International Power Generation*, December 1986/January 1987; David W. Kearney and Henry W. Price, "Overview of the SEGS Plants," presented to the Solar '87 Conference, Portland, Ore., July 1987.

70. Christopher Flavin, "Electricity from Sunlight: The Emergence of Photovoltaics," DOE, Washington, D.C., December 1984; R.L. Watts and S.A. Smith, "Photovoltaic Industry Progress from 1980 through 1986," Pacific Northwest Laboratory, Battelle Memorial Institute, Richland, Wash., June 1987; IEA, *Renewable Sources*; Terry Clausen, Chief Counsel, Solar Energy Research Institute, Golden, Colo., private communication, November 9, 1987.

71. Pre-1981 shipments and cost data from Strategies Unlimited, *1980–81 Market Review* (Mountain View, Calif.: 1981); post-1980 shipments data from Watts and Smith, "Photovoltaic Industry Progress"; post-1980 cost figures from Paul Maycock, "PV Technology, Performance, Cost, and Market Forecast to 1995," PV Energy Systems, Casanova, Va., November 1986.

72. Watts and Smith, "Photovoltaic Industry Progress"; "Communication Systems: Photovoltaics is Preferred Power Source," *ARCO News*, Summer 1986.

73. Herbert Wade, U.N. Pacific Energy Development Programme, Fiji, "The Socio-Economic Benefits of PV Applications in the Pacific," presented to the Photovoltaics: Investing in Development Conference, organized by DOE, New Orleans, La., May 4–6, 1987; "French Polynesia—World's Largest Market for Small PV Systems?" *PV News*, May 1987; William Meade, "Caribbean Project Opportunities," Renewable Energy Institute, Washington, D.C., May 1987; Richard Hansen, Enersol Associates, Somerville, Mass., private communication, September 23, 1987; IEA, *Renewable Sources*.

74. "Solarex Wins U.S. Coast Guard Contract," *Photovoltaic Insider's Report*, July 1987; Paul Maycock, presentation to Society for International Development Energy Luncheon, July 2, 1987; Roger Vielvoye, "Watching the World: Solar Power," *Oil & Gas Journal*, August 17, 1987.

75. Paul Maycock, "Consumer Products—PV's Fastest Growing Segment," *PV International*, November 1987.

76. Clausen, private communication; Watts and Smith, "Photovoltaic Industry Progress."

77. Watts and Smith, "Photovoltaic Industry Progress."

78. Ibid.; Chronar Corporation, "Energy at the Crossroads," *Annual Report '86*, Princeton, N.J., April 1987; Don Best, "PV Manufacturing Plant Would be Utility-Owned," *Renewable Energy News*, October 1985.

79. Paul Maycock, PV Energy Systems, Inc., Casanova, Va., private communication, November 17, 1987; Hubbard, Testimony before Subcommittee on Energy Research and Development.

80. "The Bad News and Good News for Photovoltaics," *Solar Today*, May/June 1987.

81. "Solarex, ARCO Solar Sue Each Other, Charging Thin Film Technology Patent Infringement," *Photovoltaic Insider's Report*, June 1987; "Chronar, Pension Fund Sign Letter of Intent on 10-MWp a-Si PV Plant," *International Solar Energy Intelligence Report*, October 20, 1987; Watts and Smith, "Photovoltaic Industry Progress."

82. IEA, *Renewable Sources*; Thomas Jaras, *Wind Energy 1987: Wind Turbine Shipments and Applications* (Great Falls, Va.: Stadia, Inc., 1987).

83. Jaras, *Wind Energy 1987*.

84. Ibid.; M.A. Illyin et al., "Pacific Gas and Electric Department of Engineering Research's Wind Energy Activities," presented at Wind Energy '87, San Francisco, Calif., October 5–8, 1987.

85. Paul Gipe, American Wind Energy Association, Tehachapi, Calif., private communications, November 5 and 6, 1987; D.R. Smith, "The Wind Farms of the Altamont Pass Area," in Annual Reviews Inc., *Annual Review of Energy, Vol. 12* (Palo Alto, Calif.: in press).

86. IEA, *Renewable Sources*; Gipe, private communication; Illyin et al., "Pacific Gas and Electric Wind Energy Activities."

87. IEA, *Renewable Sources*; Strategies Unlimited, "International Market Evaluations: Wind Energy Prospects," CEC, Sacramento, Calif., 1987.

88. R. Lynette & Assoc., Inc., "The Lessons of the California Wind Farms: How Developing Countries Can Learn From the American Experience," Redmond, Wash., 1987; Tom Gray, American Wind Energy Association, Testimony at Hearings, Subcommittee on Energy Research and Development, U.S. House of Representatives, March 11, 1987; IEA, *Renewable Sources*.

89. Average wind turbine size derived from information provided by Sam Rashkin, CEC, Sacramento, Calif., private communication, October 6, 1987; and in Paul Gipe, "An Overview of the U.S. Wind Industry," *Alternative Sources of Energy*, September/October 1985; Lynette & Assoc., "Lessons of

California"; IEA, *Renewable Sources*; Strategies Unlimited, "Wind Energy Prospects"; Kevin Porter, Renewable Energy Institute, private communications, October and November 1987.

90. Strategies Unlimited, "Wind Energy Prospects"; Strategies Unlimited, "International Energy and Trade Policies of California's Export Competitors," CEC, Sacramento, Calif., 1987.

91. Jaras, *Wind Energy 1987*.

92. Information on Danish manufacturers from Thomas Jaras, quoted in "Top Ten Listings Prove Third World Growth," *Windpower Monthly*, September 1987; Torgny Møller, Publisher, *Windpower Monthly*, Knebel, Denmark, private communication, September 9, 1987; Cathy Kramer, "The Ebeltoft Sea-Based Wind Project," *Alternative Sources of Energy*, December 1985; Strategies Unlimited, "International Energy and Trade Policies."

93. Strategies Unlimited, "Wind Energy Prospects"; IEA, *Renewable Sources*; Jaras, *Wind Energy 1987*; Costis Stambolis, "Danwin Snares Large Export Contract to Supply Windmills to Soviet Union," *International Solar Energy Intelligence Report*, September 29, 1987; "International Roundup," *International Solar Energy Intelligence Report*, September 22, 1987.

94. Strategies Unlimited, "Wind Energy Prospects"; Jaras, *Wind Energy 1987*; Gipe, "Overview of U.S. Wind Industry"; "World List of Nuclear Power Plants," *Nuclear News*, February 1986.

95. Gipe, "Overview of U.S. Wind Industry"; Lynette & Assoc., "Lessons of California"; Philip C. Cruver, "Windpower: Electrical Power Source for the Future," *SunWorld*, Vol. 11, No. 3, 1987; Hubbard, Testimony before Subcommittee on Energy Research and Development.

96. IEA, *Energy Policies and Programmes of IEA Countries: 1986 Review* (Paris: OECD, 1987).

97. Ibid.; Population Reference Bureau, *1986 World Population Data Sheet* (Washington, D.C.: 1986); World Bank, *World Development Report 1987* (New York: Oxford University Press, 1987).

98. Sweden's per capita R&D spending derived from IEA, *Policies and Programmes*, and from Population Reference Bureau, *1986 World Population Data Sheet*; information on Swedish biomass programs from IEA, *Policies and Pro-*

grammes; from "Green Power: Biofuels are a Growing Concern," *Scientific American*, August 1984; and from Allerd Stikker, President, Trans-form Foundation, London, private communication, September 28, 1987; information on Sweden's nuclear policy from "Swedish Plan for Nuclear Phase-Out," *European Energy Report*, May 29, 1987.

99. Strategies Unlimited, "International Energy and Trade Policies"; "Swiss Program Reflects Interest in Indigenous, Non-Polluting Energy," *Solar Update*, May 1987; Strategies Unlimited, "Wind Energy Prospects"; IEA, *Policies and Programmes*; Judith Perera, "Indian Government Draws Up Plans to Exploit Renewable Energy," *Solar Energy Intelligence Report*, August 11, 1987.

100. Strategies Unlimited, "International Energy and Trade Policies."

101. IEA, *Renewable Sources*; Ronald DiPippo, Southeastern Massachusetts University, North Dartmouth, Mass., private communication, June 29, 1987; Ronald DiPippo, "Geothermal Power Plants, Worldwide Status—1986," *Geothermal Resources Council Bulletin*, December 1986; 1991 capacity figure and Electric Power Research Institute study cited in Donald Finn, "Expanding Geothermal Industry," *Alternative Sources of Energy*, October 1987.

102. Arun Madan, "Amorphous Silicon: From Promise to Practice," *IEEE Spectrum*, September 1986.

103. Jean Jarecki, Federal Energy Regulatory Commission (FERC), Washington, D.C., private communication, September 30, 1987; George D. Rodriguez, "Edison Embarks on Another First: The Chino 10 MW Battery Energy Storage Project," *Research Newsletter*, Southern California Edison Company, Third Quarter 1987.

104. Requested capacity figure a Worldwatch estimate based upon figures provided by Charles Reeder, FERC, private communication, October 6, 1987; and in *Cogeneration and Small Power Monthly*, September 1987; installed capacity from Edison Electric Institute, "1985 Capacity and Generation Non-Utility Sources"; total contribution from IEA, *Renewable Sources*; Sweden's and Portugal's total energy supply from British Petroleum, *BP Statistical Review of World Energy* (London: 1987).

105. Share of proposed generating capacity that would rely on renewables based upon FERC, "The Qualifying Facilities Report," Washington, D.C., January 1, 1987; on Reeder, private communication, and on *Cogeneration & Small Power Monthly*, July, August, and September 1987. For an overview of he debate about competitive bidding by utilities and small power producers, ee *Alternative Sources of Energy*, May/June 1987.

06. CEC, "Relative Cost."

07. For information on Greek legislation, see "Country Profile Greece,")PEC Bulletin, April 1987; New Zealand information from David Pate, Energy ;onsultant, private communication, November 30, 1987; on proposed legisla-ion in Denmark, Israel, and West Germany, see Strategies Unlimited, "International Energy and Trade Policies"; on India, see Marie D'Souza, "Energy Banking Finally Underway In India," *Windpower Monthly*, May 1987; on the Netherlands, see Jos Van Beek, "Second Dutch Firm Home and Dry in Subsidy Race," *Windpower Monthly*, August 1987; on Pakistan, see "Sustainable Surroundings," *Horizons*, Spring 87; information on Turkey from Porter, private communication; on the United Kingdom, see Andrew Holmes et al., *Power on the Market: Strategies for Privatising the UK Electricity Industry* (London: Financial Times Business Information, 1987); on Sweden, see "Swedish Plan for Nuclear Phase-Out"; on Nepal, see AID, *Decentralized Hydropower*; on China, see Eugene Chang, "Little Plants Give Lots of Power," *China Daily*, December 14, 1985.

108. International Co-operative Programme, "Forest Damage and Air Pollution."

109. José Goldemberg et al., *Energy for a Sustainable World* (Washington, D.C.: World Resources Institute, 1987).

110. Mark Kosmo, *Money to Burn? The High Costs of Energy Subsidies* (Washington, D.C.: World Resources Institute, 1987). Note that China's subsidies cannot simply be added because substantial double counting would occur. Amory B. Lovins and L. Hunter Lovins, "Oil Wells and Battleships are the Answer! (But What Was the Question?)" (draft), Rocky Mountain Institute, Old Snowmass, Colo., August 13, 1987.

111. Stephen W. Sawyer, *Renewable Energy: Progress, Prospects* (Washington, D.C.: Association of American Geographers, 1986); Strategies Unlimited, "International Energy and Trade Policies"; Porter, private communication; Van Beek, "Second Dutch firm"; Smith, "The Wind Farms of Altamont Pass."

112. World Commission on Environment and Development, *Our Common Future*; Hall, *Biomass for Energy*; Goldemberg et al., *Energy for Development*; Larson et al., "Steam-Injected Gas-Turbine Cogeneration."

113. Ross Pumfrey et al., "The Philippines Trade and Investment Laws Rela ing to Renewable Energy," Renewable Energy Institute, Washington, D.C March 1987.

114. DiPippo, "Geothermal Power Plants"; Pumfrey et al., "Philippines Trad and Investment."

115. Ross Pumfrey et al., "India Trade and Investment Laws Relating t Renewable Energy," Renewable Energy Institute, Washington, D.C., Marc 1987; Perera, "Indian Government Draws Up Plans"; "World List of Nucle Power Plants."

116. IEA, *Policies and Programmes;* Strategies Unlimited, "International Energ and Trade Policies"; Scott Sklar, "International Trade Policy for the Renewab Energy Industries: An Assessment," *Solar Today*, March/April 1987.

CYNTHIA POLLOCK SHEA is a Senior Researcher with Worldwatc Institute, and coauthor of *State of the World 1988*. Her research deal with energy and waste management technologies and policies. She i a graduate of the University of Wisconsin-Madison, where she stud ied economics, environmental studies, and international relations.

THE WORLDWATCH PAPER SERIES

No. of Copies*

______ 1. **The Other Energy Crisis: Firewood** by Erik Eckholm.
______ 3. **Women in Politics: A Global Review** by Kathleen Newland.
______ 7. **The Unfinished Assignment: Equal Education for Women** by Patricia L. McGrath.
______ 10. **Health: The Family Planning Factor** by Erik Eckholm and Kathleen Newland.
______ 13. **Spreading Deserts—The Hand of Man** by Erik Eckholm and Lester R. Brown.
______ 16. **Women and Population Growth: Choice Beyond Childbearing** by Kathleen Newland.
______ 18. **Cutting Tobacco's Toll** by Erik Eckholm.
______ 21. **Soft Technologies, Hard Choices** by Colin Norman.
______ 25. **Worker Participation—Productivity and the Quality of Work Life** by Bruce Stokes.
______ 28. **Global Employment and Economic Justice: The Policy Challenge** by Kathleen Newland.
______ 29. **Resource Trends and Population Policy: A Time for Reassessment** by Lester R. Brown.
______ 30. **The Dispossessed of the Earth: Land Reform and Sustainable Development** by Erik Eckholm.
______ 31. **Knowledge and Power: The Global Research and Development Budget** by Colin Norman.
______ 33. **International Migration: The Search for Work** by Kathleen Newland.
______ 34. **Inflation: The Rising Cost of Living on a Small Planet** by Robert Fuller.
______ 35. **Food or Fuel: New Competition for the World's Cropland** by Lester R. Brown.
______ 36. **The Future of Synthetic Materials: The Petroleum Connection** by Christopher Flavin.
______ 37. **Women, Men, and The Division of Labor** by Kathleen Newland.
______ 38. **City Limits: Emerging Constraints on Urban Growth** by Kathleen Newland.
______ 39. **Microelectronics at Work: Productivity and Jobs in the World Economy** by Colin Norman.
______ 40. **Energy and Architecture: The Solar and Conservation Potential** by Christopher Flavin.
______ 41. **Men and Family Planning** by Bruce Stokes.
______ 42. **Wood: An Ancient Fuel with a New Future** by Nigel Smith.
______ 43. **Refugees: The New International Politics of Displacement** by Kathleen Newland.
______ 44. **Rivers of Energy: The Hydropower Potential** by Daniel Deudney.
______ 45. **Wind Power: A Turning Point** by Christopher Flavin.
______ 46. **Global Housing Prospects: The Resource Constraints** by Bruce Stokes.
______ 47. **Infant Mortality and the Health of Societies** by Kathleen Newland.
______ 48. **Six Steps to a Sustainable Society** by Lester R. Brown and Pamela Shaw.
______ 49. **Productivity: The New Economic Context** by Kathleen Newland.
______ 50. **Space: The High Frontier in Perspective** by Daniel Deudney.
______ 51. **U.S. and Soviet Agriculture: The Shifting Balance of Power** by Lester R. Brown.
______ 52. **Electricity from Sunlight: The Future of Photovoltaics** by Christopher Flavin.

*Worldwatch Papers 2, 4, 5, 6, 8, 9, 11, 12, 14, 15, 17, 19, 20, 22, 23, 24, 26, 27, and 32 are out of print.

_____ 53. **Population Policies for a New Economic Era** by Lester R. Brown.
_____ 54. **Promoting Population Stabilization** by Judith Jacobsen.
_____ 55. **Whole Earth Security: A Geopolitics of Peace** by Daniel Deudney.
_____ 56. **Materials Recycling: The Virtue of Necessity** by William U. Chandler.
_____ 57. **Nuclear Power: The Market Test** by Christopher Flavin.
_____ 58. **Air Pollution, Acid Rain, and the Future of Forests** by Sandra Postel.
_____ 59. **Improving World Health: A Least Cost Strategy** by William U. Chandler.
_____ 60. **Soil Erosion: Quiet Crisis in the World Economy** by Lester Brown and Edward Wolf.
_____ 61. **Electricity's Future: The Shift to Efficiency and Small-Scale Power** by Christopher Flavin.
_____ 62. **Water: Rethinking Management in an Age of Scarcity** by Sandra Postel.
_____ 63. **Energy Productivity: Key to Environmental Protection and Economic Progress** by William U. Chandler.
_____ 64. **Investing in Children** by William U. Chandler.
_____ 65. **Reversing Africa's Decline** by Lester Brown and Edward Wolf.
_____ 66. **World Oil: Coping With the Dangers of Success** by Christopher Flavin.
_____ 67. **Conserving Water: The Untapped Alternative** by Sandra Postel.
_____ 68. **Banishing Tobacco** by William U. Chandler.
_____ 69. **Decommissioning: Nuclear Power's Missing Link** by Cynthia Pollock.
_____ 70. **Electricity For A Developing World** by Christopher Flavin.
_____ 71. **Altering the Earth's Chemistry: Assessing the Risks** by Sandra Postel.
_____ 72. **The Changing Role of the Market in National Economies** by William U. Chandler.
_____ 73. **Beyond the Green Revolution: New Approaches for Third World Agriculture** by Edward C. Wolf.
_____ 74. **Our Demographically Divided World** by Lester R. Brown and Jodi L. Jacobson.
_____ 75. **Reassessing Nuclear Power: The Fallout From Chernobyl** by Christopher Flavin.
_____ 76. **Mining Urban Wastes: The Potential for Recycling** by Cynthia Pollock.
_____ 77. **The Future of Urbanization: Facing the Ecological and Economic Constraints** by Lester R. Brown and Jodi L. Jacobson.
_____ 78. **On the Brink of Extinction: Conserving Biological Diversity** by Edward C. Wolf.
_____ 79. **Defusing the Toxics Threat: Controlling Pesticides and Industrial Waste** by Sandra Postel.
_____ 80. **Planning the Global Family** by Jodi L. Jacobson.
_____ 81. **Renewable Energy: Today's Contribution, Tomorrow's Promise** by Cynthia Pollock Shea.

_____ **Total Copies**

Bulk Copies (any combination of titles) **Single Copy** $4.00
2-5: $3.00 each 6-20: $2.00 each 21 or more: $1.00 each

Calendar Year Subscription (1988 subscription begins with Paper 81) U.S. $25.00 _____

Make check payable to Worldwatch Institute
1776 Massachusetts Avenue NW, Washington, D.C. 20036 USA

Enclosed is my check for U.S. $ _____

____
name
____
address
____
city **state** **zip/country**